Big Data Driven Supply Chain Management

A Framework for Implementing Analytics and Turning Information into Intelligence

Nada R. Sanders, Ph.D.

Distinguished Professor of Supply Chain Management
D'Amore-McKim School of Business
Northeastern University

Associate Publisher: Amy Neidlinger
Executive Editor: Jeanne Glasser Levine
Development Editor: Natasha Torres
Operations Specialist: Jodi Kemper
Cover Designer: Alan Clements
Managing Editor: Kristy Hart
Senior Project Editor: Lori Lyons
Copy Editor: Karen Annett
Proofreader: Debbie Williams
Indexer: Erika Millen
Senior Compositor: Gloria Schurick
Manufacturing Buyer: Dan Uhrig

© 2014 by Nada R. Sanders
Pearson Education, Inc.
Upper Saddle River, New Jersey 07458

For information about buying this title in bulk quantities, or for special sales opportunities (which may include electronic versions; custom cover designs; and content particular to your business, training goals, marketing focus, or branding interests), please contact our corporate sales department at corpsales@pearsoned.com or (800) 382-3419.

For government sales inquiries, please contact governmentsales@pearsoned.com.

For questions about sales outside the U.S., please contact international@pearsoned.com.

Company and product names mentioned herein are the trademarks or registered trademarks of their respective owners.

Printed in the United States of America

First Printing June 2014

ISBN-10: 0-13-380128-4
ISBN-13: 978-0-13-380128-6

Pearson Education LTD.
Pearson Education Australia PTY, Limited.
Pearson Education Singapore, Pte. Ltd.
Pearson Education Asia, Ltd.
Pearson Education Canada, Ltd.
Pearson Educación de Mexico, S.A. de C.V.
Pearson Education—Japan
Pearson Education Malaysia, Pte. Ltd.

Library of Congress Control Number: 2014934904

This book is dedicated to the reader.
With knowledge and understanding,
we can make our enterprising efforts more
efficient, effective, and intelligent.

Contents

Table of Contents

Foreword

The surge of professional and academic interest in the topic of big data analytics has been similar to the gold rush—there is much commotion but few know where to look for the real payoff. Although it has long been obvious that utilizing emerging information technology is vital to remaining competitive, the nature of that technology has shifted in a tectonic way, and thought leadership must keep pace.

Many companies have yet to leverage big data analytics to transform their supply chain operations. Many are awash in data but are unsure how to use it to drive their supply chains. Many are engaging in fragmented utilization or implementation rather than a systematic and coordinated effort. The results are isolated benefits, lack of insight and competitiveness, and supply chains plagued with inefficiencies and cost overruns. Others are unsure how to even begin, particularly small and medium-sized firms.

Why all the confusion? The reason is that companies lack a clear roadmap for how to implement big data analytics in a meaningful and cost-effective manner.

This book attempts to remedy the situation by providing a systematic framework for companies on *how* to implement big data analytics across the supply chain to turn information into intelligence and achieve a competitive advantage. This end-to-end perspective on the application of big data analytics provides a much-needed conceptual organization to this topic while linking strategy with tactics. Furthermore, this roadmap shows organizational leaders *how* to implement the type of organizational change big data analytics requires.

Acknowledgments

I am most grateful to my family—especially my parents, who taught me how to create order out of chaos, and convey with clarity.

I would also like to thank all the professionals who shared their experiences and provided insights into this work. Special thanks goes to Jeanne Glasser Levine at Pearson, as well as the other talented staff at Pearson who have contributed work to this book.

About the Author

Nada R. Sanders, Ph.D., is Distinguished Professor of Supply Chain Management at the D'Amore-McKim School of Business at Northeastern University, and she holds a Ph.D. from the Ohio State University. She is an internationally recognized thought leader and expert in forecasting and supply chain management. She is author of the book *Supply Chain Management: A Global Perspective* and is coauthor of the book *Operations Management,* in its 5th edition. She was ranked in the top 8 percent of individuals in the field of operations management from a pool of 738 authors and 237 different schools by a study of research productivity in U.S. business schools. Nada is a Fellow of the Decision Sciences Institute, and has served on the Board of Directors of the International Institute of Forecasters (IIF), Decision Sciences Institute (DSI), and the Production Operations Management Society (POMS). Her research focuses on the most effective ways for organizations to use technology to achieve a competitive advantage.

Part I

"Big" Data Driven Supply Chains

1

A Game Changer

The era of radically different competition is here. It is a tsunami that has transformed entire industries and left numerous casualties in its wake. Like Gutenberg's invention of the printing press changing the world through printing, the move toward big data is creating an equally tectonic shift in business and society. Transform or be left behind.

Consider the fate of Borders.[1] In 1971, the company opened its first store in Ann Arbor, Michigan, when the book industry was a different place. In 2011, 40 years later, the bookstore chain closed its doors. So, what happened? Borders fell behind the curve on embracing the Web and the digital world of data. Not understanding that the rules of the game had changed, Borders had outsourced its online bookselling to Amazon.com. So any time you visited Borders.com, you were redirected to Amazon. Playing by the old rules made this seem like a smart decision. In the new world, however, there was a problem.

To jump on the tails of Amazon and leverage its competitive priorities did not take into account that playing in the digital world was *the* competitive priority. Relinquishing control to another company would simply cut into the company's customer base. Also, not understanding that the world was now a digital place, Borders did not embrace e-books, like Amazon and Barnes & Noble. Walking into Borders was like walking into a bookstore of yesteryear. The outcome was predictable.

The competitive world Borders lived in was one where booksellers tracked which books sold and which did not. Loyalty programs could help tie purchases to individual customers. That was about it. Then shopping moved online. The ability to understand and track customers changed dramatically. Online retailers could track every aspect of what customers bought. They could track what customers looked at, how they navigated through the site, how long they hovered over a site, and how they were influenced by promotions and page layouts. They were now able to develop microsegments of individual customers and groups based on endless characteristics. They could then create individually targeted promotions. Then algorithms were developed to predict what books individual customers would like to read next. These algorithms were self-teaching and performed better every time the customer responded to a recommendation. Traditional retailers like Borders simply couldn't access this kind of information. They could not compete in a timely manner.

And Amazon? With its Kindle e-book readers and convincing hundreds of publishers to release their books on the Kindle format, the company has cornered the market. It has "datafied" books—turning them into quantified format that can be tabulated and analyzed.[2] This allows Amazon everything from recommending books to using algorithms to find links among the topics of books that might not otherwise be apparent. Embracing the digital age, technology, and data-driven decisions, the company is moving well beyond wanting to be the biggest bookstore on the Internet. It is moving toward being the most dominant retailer in the world. Amazon understands that this means using big data and technology to manage its entire supply chain in a synchronized manner. In fact, Jeff Bezos, Amazon's CEO, is known for demanding rigorous quantification of customer reactions before rolling out new features.[3] Data and technology have been used to coordinate everything from customer orders to fulfillment, inventory management, labor, warehousing, transportation, and delivery.

Amazon is not the only one. Leading-edge companies across the globe have scored successes in their use of big data. Consider Walmart, Zara, UPS, Tesco, Harrah's, Progressive Insurance, Capital One, Google, and eBay.[4] These companies have succeeded in this game-changing environment by embracing and leading the change. They have used big data analytics to extract new insights and create new forms of value in ways that have changed markets, organizations, and business relationships.

1.1 Big Data Basics

To fully understand the impact of big data analytics, we first need to have a clear idea of what it actually is. In this section we explain big data basics. We define the key concepts of big data analytics and explain how these concepts find novel applications across business. This will set up the book's subsequent discussions of big data analytics applications in supply chain management.

1.1.1 Big Data

Big data is simply lots of data. The term *big data* specifically refers to large data sets whose size is so large that the quantity can no longer fit into the memory that computers use for processing. This data can be captured, stored, communicated, aggregated, and analyzed.[5] There is no specific definition of the size of big data, such as the number of terabytes or gigabytes. The reason is that this is a moving target. Technology is advancing over time and the size of data sets that are considered big data will also increase.

As the volume of data has grown, so has the need to revamp the tools used for analyzing it. That is how new processing technologies like Google's MapReduce and its open source equivalent, Hadoop, were developed. These new technologies enable companies to manage far-larger quantities of data than before. Most important, unlike in the past, this data does not need to be placed in neat rows and columns as traditional data sets to be analyzed by today's technology.

Big data comes in different forms. It includes all kinds of data from every source imaginable. It can be structured or unstructured. It can be a numerical sequence or voice and text and conversation. It can come in the form of point-of-sale (POS), radio-frequency identification (RFID), or Global Positioning System (GPS) data, or it can be in the form of Twitter feeds, Facebook, call centers, or consumer blogs. Today's advanced analytical tools allow us to extract meaning from all types of data.

1.1.2 Analytics

Analytics is applying math and statistics to these large quantities of data. When we apply math and statistics to big data—often called *big data analytics*—we can gain insights into the world around us unlike ever before. We can infer probabilities or likelihoods that something will happen.

We are used to this in our everyday life. We are accustomed to e-mail filters that estimate the likelihood that an e-mail message is spam or that the typed letters *teh* are supposed to be *the*. The key is that these systems perform well because they are fed with lots of data on which to base their predictions. Moreover, the systems are built to improve themselves over time, by keeping tabs on the best signals and patterns to look for as more data is fed in. Think about "teaching" your e-mail filter that a type of e-mail is a spam by labeling similar e-mails.

It is through big data that Walmart learned that customers prefer to stock up on the sugary treat Pop-Tarts during a hurricane,[6] eBay identified which Web designs generate the highest sales,[7] and Progressive Insurance learned how to optimize insurance premiums by risk category.[8]

Even small companies have benefited. Consider the online music equipment retailer The Musician's Friend. Using basic analytics, the company was able to compare different versions of its Web page to identify customer preferences. The preferred site generated a 35 percent increase in sales over the original home page. This simple change resulted in a measurable improvement on return on investment (ROI).[9]

1.1.3 Big Data and Analytics: The Perfect Duo

To set the record straight, big data without analytics is just lots of data. We've been accumulating a lot of data for years. Analytics without big data is simply mathematical and statistical tools and applications. Tools such as correlation and regression, for example, have been around for decades. In fact, Google's director of research, Peter Norvig, explained it well by saying: "We don't have better algorithms. We just have more data."[10]

However, it is the combination that makes the difference. It is through the combination of big data and analytics that we can get the really meaningful insights and turn information into business intelligence (see Figure 1.1) Also, big data and analytics build on each other. Continued application of even simple analytical tools results in their improvement, refinement, and sophistication. Consider that as you increasingly identify the number of e-mails as spam, the filter "learns" and becomes better at correctly identifying spam. It is for this reason we use the term *big data analytics* throughout this book to refer to the application of analytics to these large data sets.

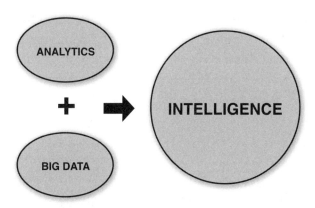

Figure 1.1 Turning information into intelligence

1.1.4 New Computing Power

How can companies extract intelligence out of these huge amounts of data? This is made possible through today's massive computing power available at a lower cost than ever before. Large data, coupled with larger and more affordable computing power, means that you can do on a larger scale that which cannot be done on a smaller one. Improvements in computing have resulted in large advances in capability. This has enabled high-level analytics to be performed on these large and unstructured data sets.

Processing power has increased over the years just as predicted by Moore's law.[11] The law is named after Intel cofounder Gordon E. Moore and states that the amount of computing power that can be purchased for the same amount of money doubles about every two years. This law has proven correct. We have seen computers becoming faster and memory more abundant. Similarly, storage space has expanded through *cloud computing*, which refers to the ability to access highly scalable computing resources through the Internet. Cloud computing is often available at a lower cost than that required for installation on in-house computers. This is because resources are shared across many users. Further, the performance of the algorithms that drive so many of our systems has also increased. Therefore, the gains from big data are a combination of the size of current data sets coupled with rapidly increasing processing capability and improved algorithms.

Hadoop is an open source technology platform that has received a great deal of buzz as it was designed to solve problems with lots of data. In fact, Hadoop was specifically designed to deal with big data that is a mixture of complex and structured data that does not fit nicely into tables. It was originated by Google for its own use for indexing the Web and examining user behavior to improve performance algorithms. Yahoo! then furthered its development for enterprise purposes. Hadoop uses distributed applications across many servers. The database is distributed over a large number of machines. Spreading data over multiple machines greatly improves computing capability. Because the tables are divided and distributed into multiple servers, the total number of rows in each table in each database is reduced. This reduces index size and substantially improves search performance. The database is typically divided into partitions called *database shards*. A database shard can be placed on separate hardware and multiple shards can be placed on multiple machines. Database shards significantly improve performance. The segment of the database placed on a shard can be based on real-world segmentation. This can greatly help analysis—such as separating Canadian customers versus American customers. This makes it especially easy to query a particular segment of the data or evaluate comparisons across segments.

1.1.5 New Problem Solving

Just a few years ago, the topic of analytics and computing power would have concerned only a few data geeks. Today, however, big data is an imperative for all business leaders across every industry and sector—from health care to manufacturing. The ability to capture, store, aggregate, and combine data—and then perform deep analyses—has now become accessible to virtually all organizations. This will continue as costs of computing power, digital storage, and cloud computing continue to drop. These advancements will further break down technology barriers and level the playing field, especially for small and medium-sized firms. Just consider that today an individual can purchase a disk drive with the capacity to store all of the world's music for less than $600.[12] In fact, the cost of storing a terabyte of data has fallen from $1 million in the 1970s to $50 today.[13]

What does this mean for business? Simply put, increasingly sophisticated analytical techniques combined with growing computer horsepower mean extracting unparalleled business insights. It means new and revolutionary problem-solving capability. Big data is not about the data itself. It is about the ability to solve problems better than ever before. And now pretty much everyone can do it.

1.2 What Is Different?

Companies have been capturing data for years and conducting analysis to gain market intelligence. It is natural to wonder what exactly is different today.

The difference is *scale*. This scale is in terms of the amount of data *and* the computing capability to analyze it. Combined, these elements can offer objective, evidence-based insights into virtually every aspect of a business. This has created a new level of competitiveness and the opportunity for companies to use this intelligence as a competitive advantage. We are living through a data explosion. According to a recent *New York Times* article, "Data is a vital raw material of the information economy, much as coal and iron ore were in the Industrial Revolution. But the business world is just beginning to learn how to process it all."[14]

The three characteristics of big data—volume, velocity, and variety—are what make big data different from all the other data collected in the past (see Figure 1.2).[15]

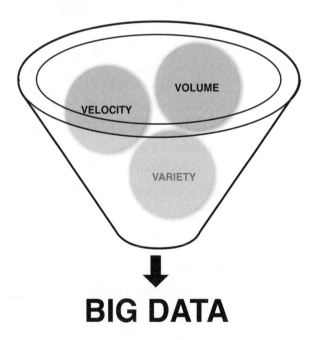

Figure 1.2 The three Vs of big data

1.2.1 Volume

Today's data is huge and data is everywhere. Consider that Google receives more than three billion queries every day, a volume that is thousands of times the quantity of all printed material in the U.S. Library of Congress; Facebook gets more than 10 million new photos uploaded every hour; Walmart conducts a million transactions per hour; and the New York Stock Exchange (NYSE) trades 490 million shares per day.[16] The amount of data generated will continue to grow exponentially. In fact, the number of RFID tags sold globally is projected to rise from 12 million in 2011 to 209 billion in 2021.[17]

Manufacturers and retailers are collecting data all along their supply chains. This includes data from POS, GPS, and RFID data, to data emitted by equipment sensors, to social media feeds. Virtually all companies have information technology (IT) systems in their back offices. The world we live in is enveloped in data. When converting terabytes and exabytes into meaningful terms, it is estimated that the data that companies and individuals are producing and storing is equivalent to filling more than 60,000 U.S. Libraries of Congress.[18] Where is this data going? It is accumulating in large pools growing larger by the minute. This is *big data*. In fact, many companies do not even recognize the data they possess and how valuable it is. Data can be traded or sold, and it has economic value. Data is the new asset.

1.2.2 Velocity

Yes, the data is large. The data is growing. And it is growing rapidly.

Data has become a deluge flowing into every aspect of business and everyday life. Companies are capturing exponentially growing volumes of transactional data. They are also capturing volumes of information about their customers, suppliers, and operations. Consider that there are millions of sensors embedded in physical devices all around us, including mobile phones, smart energy meters, automobiles, and industrial machines. These devices capture and communicate data in what is called the age of the *Internet of Things*.[19]

Companies and individuals are generating a tremendous amount of digital *exhaust data.* This is data that is created as a by-product of other activities and is generated just by going about everyday business. Consumers going about their day through texting, communicating, browsing, searching, and buying are creating *digital trails.* These trails can be captured, monitored, and analyzed. Social media sites, smartphones, PCs, and laptops have enabled billions of individuals around the world to continue adding to the amount of big data available. The growing amount of multimedia content has played a major role in the exponential growth in the amount of big data. In fact, each second of high-definition video, for example, generates more than 2,000 times as many bytes as required to store a single page of text.[20]

Companies are collecting data with increasingly greater granularity and frequency. They can capture every customer transaction. They can attach increasingly personal information to these transactions. As a result, they are collecting more information about consumer behavior in many different environments. Companies can then aggregate and disaggregate this data in infinite combinations to mathematically optimize and determine the best marketing strategies, custom-tailored shopping experiences, flexible and targeted product designs, and an optimized fulfillment system.

1.2.3 Variety

Big data is also in every form imaginable. Most of us think of data as numbers neatly stored in columns and rows. However, big data is in combinations of forms and this variety is growing. It is in the form of structured data that we are familiar with but may also be readings from sensors, GPS signals from cell phones, and POS and RFID data. It is also unstructured, such as text and voice messages, social networks feeds, and blogs.

Sources of big data are everywhere—from sensors such as RFID and POS data at retail checkouts, to geolocation information transmitted from GPS signals, to vibration and heat sensors on equipment, to social media feeds (think of Twitter "re-tweets" and Facebook "likes"), to maintenance logs, to customer complaints (see Figure 1.3). Most smartphones have GPS capabilities that enable tracking. Cars alone are stuffed with chips, sensors, and software that upload performance data to the carmakers' computers when the vehicle is serviced. Typical mid-tier vehicles now have some 40 microprocessors and electronics that account for one third of their cost.[21] How else does your car manufacturer inform you that it is time for service?

Figure 1.3 Sources of data are everywhere

1.3 What Does It Mean?

The bottom line is that big data analytics enables converting information into an unprecedented amount of business intelligence. It allows companies to precisely understand *what* happened in the past and *why*, and better predict the future. The result is a superior competitive capability.

1.3.1 Big Business Intelligence

Not using big data analytics in today's business world is akin to making deliveries on horseback while competitors are using a truck. Or using carrier pigeons while everyone else is using airfreight. A company just can't compete without it.

What can big data analytics tell us?

The size of data makes it possible to spot connections and details in the data that are otherwise impossible to spot. Granular analysis of subcategories and submarkets is enabling the understanding of customers and markets unlike ever before. A task that used to be accomplished based on traditional marketing tools—such as focus groups and surveys—to try to determine what the customer wants is now computed based on scientific methods. We no longer guess or use hunches. We know exactly what each customer wants to buy. This changes the way we sell products. This information also drives entire supply chains.

The large amounts of data enable the establishments of *norms* in the data. This also helps identify data that is outside of the norms— namely outliers. This technique, for example, is what we see today in the identification of credit card fraud. The algorithm automatically detects a change from the norm. This is only possible when there is a large amount of data. This has huge applications any time we are looking for a deviation. This is what UPS does, for example, when it monitors its vehicle fleets for preventive maintenance.[22] The company cannot afford a breakdown in its fleet of vehicles. Before big data analytics, the company routinely replaced parts, but that was wasteful. Then in 2000, the company embarked on a program of using sensors to capture data on vehicle performance and notice deviations that indicate a time to intervene before a breakdown.

What else can big data analytics do?

1.3.2 Predicting the Future

Predictive analytics uses a variety of techniques—such as statistics, modeling, and data mining—to analyze current and historical facts to make predictions about the future.[23] This is one of the most significant aspects of big data analytics. It is the ability to foresee events before they happen by sensing small changes over time. For example, IBM's Watson computer uses an algorithm to predict best medical treatments[24] and UPS uses analytics to predict vehicle breakdown.[25] By placing sensors on machinery, motors, or infrastructure like bridges, we can monitor the data patterns they give off, such as heat, vibration, stress, and sound. These sensors can detect changes that may indicate looming problems ahead—essentially forecasting a problem.

Things do not break down all at once. There is a gradual wear and tear over time. In the past, our technology, sensors, and analytics were not sophisticated enough to detect these changes. Today, armed with sensor data, correlation analysis, and similar methods, we can identify the specific patterns that typically crop up before something breaks. This may be the sound of a motor, excessive heat from an engine, or vibration from the bridge. In health care, it may be changes in a patient's vitals before the onset of disease. Google is famous for identifying location and propagation of the flu by simply tracking the volume and type of queries in its search engine.[26]

1.3.3 Fewer Black Swans

Black swans is a term used to describe high-impact, low-probability events.[27] Historically, we assumed that these could not be predicted. However, with big data analytics, that is rapidly changing. With big data analytics, the number of events that we used to consider unpredictable and purely random is getting smaller. We are now able to identify and spot changes in systems that indicate potential failure. Just consider the accuracy of the prediction of hurricane Sandy provided by the NOAA weather satellite. Only a few years earlier, this type of event would have been considered a black swan.

Spotting the abnormality early on enables the system to send out a warning so that a new part can be installed, preparation before an impending tsunami can be made, or the malfunction can be fixed. The aim is to identify and then watch a good proxy for the event we are trying to forecast, and thereby predict the future. External events—such as weather, traffic, or road construction—can be tracked and the supply chain can respond. Traffic can be rerouted or knowledge of outbreaks of flu can be used to determine which areas may need more supplies of ibuprofen, chicken soup, or cough drops. This ability is a game changer for risk management.

1.3.4 Explain **What** *Has Happened*

One of the most powerful analytics tools we can use on big data is *correlation analysis.* Correlation analysis has been used for decades. What is different today are the insights obtained when applied to the huge amounts of data. Correlations tell us whether there is a relationship between any set of variables. It doesn't tell us why there is a relationship. In the world of statistics, it is "quick and dirty" but offers important insights.

Correlations let us analyze a phenomenon by identifying a useful proxy for it. The idea is that if A often takes place together with B, we need to monitor B to predict that A will happen.

Consider the case of Target and identification of pregnant customers.[28] Big data analytics was able to identify the precise purchasing bundle associated with a female customer in the second trimester of pregnancy. Those who have seen the highly publicized story might recall the father of a 16-year-old girl who was very angry at Target for sending his daughter baby coupons—only to discover that indeed she was pregnant. The analytics perfectly targeted her—no pun intended.

Correlation analysis also points the way for causal investigations, by telling us which two things are potentially connected. This then tells us where to investigate further. This provides information on where to go into modeling, causation, and optimization. This is an important benefit. It points us in the right direction and enables us to know where to dig deep with more sophisticated analytics applications such as supply chain optimization.

1.3.5 Explain **Why** *Things Happen*

Correlation analysis tells whether a relationship exists. More advanced statistical applications enable us to go beyond understanding whether a relationship exists and delve deeper into understanding causations.

1.3.5a Supply Chain Optimization

Supply chain optimization is the application of mathematical and statistical tools to develop optimal solutions to supply chain problems. This enables analysts to create models to simulate, explore contingencies, and optimize supply chains. Many of these approaches employ some form of linear programming software and solvers. This allows the program to maximize a particular goal given a set of variables and constraints. This includes the optimal placement of inventory within the supply chain, minimizing the carbon footprint or minimizing operating costs, such as manufacturing, transportation, and distribution costs.

1.3.5b Randomized Testing

Big data and analytics have enabled companies to use randomized testing to conduct experiments to "test and learn"—sometimes called *design of experiments*. Randomized testing is a statistical method that involves conducting, analyzing, and interpreting tests to evaluate which factors impact variables of interest. For example, this might be asking whether planned changes in delivery or store layouts will increase customer purchases. Randomized testing is at the heart of the scientific method. Without random assignment to test groups, and without a control group, it is impossible to know which improvements are actually due to the changes being made. This type of large-scale testing is now possible as there is lots of data to compare and analyze.

Another significant enabler is that many current software applications are designed for people with little statistical training. New software makes it possible to conduct design of experiments by businesspeople rather than professional statisticians. For example, testing alternative versions of Web sites is relatively straightforward. This type of testing is simple and is becoming widely practiced in online retailers. Simple A/B experiments, such as comparing two versions of a Web site—A versus B—can be easily structured. The online retailer eBay, for example, routinely conducts experiments with different aspects of its Web site.[29] The site generates huge amounts of data as there are more than a billion page views per day. This enables eBay to conduct multiple experiments concurrently and not run out of treatment and control groups. Similarly, the North Carolina food retailer Food Lion uses testing to try out new retailing approaches—again simply comparing A versus B.[30] This ranges from comparing new store formats to simple tactical decisions.

1.4 Transformations

1.4.1 Business Ramifications

Consider the following examples of companies that have implemented big data analytics:

- The global cement giant CEMEX has successfully applied analytics to its distinctive capability of optimized supply chains and delivery times.[31]

- Walmart relies extensively on analytics to run its entire supply chain.

- At Deere & Company, a new way of optimizing inventory saved the company $1.2 billion in inventory costs between 2000 and 2005.[32]

- Proctor & Gamble used operations research methods to reorganize sourcing and distribution approaches in the mid-1990s and saved the company $200 million in costs.[33]

- Amazon claims its latest advanced analytics can now predict purchases *before* they happen. Based on the pattern of customer computer searches and how long the cursor lingers over a Web site, the company plans to start bundling and shipping items to distribution centers in advance of actual purchases.[34]

Questions that were once based on intuition and guesswork can now be answered in objective and quantifiable terms. Big data analytics answers business questions such as the following:

- What does the future look like? What do our customers want?
- What is the reason for our success? Is our strategy working?
- What activities should we pursue in the future? Which resources should we invest in?
- What do we do to minimize our risk exposure? How do we protect ourselves from business disruptions?

1.4.2 Changing the Present and Future

The ability to answer these questions changes virtually every aspect of business. It enables understanding both the present and future. As such, it can enhance a company's competitive position by better predicting competition and markets. It can dramatically improve operational and supply chain performance. For companies that have implemented big data analytics, it can increase productivity and improve efficiency, quality, and preventive maintenance. It can help manage suppliers and customers, as well as logistics and transportation operations. It can also better evaluate strategy, improve forecasting, help prepare for disruptions, and, overall, improve risk management.

Harnessing big data analytics has the potential to improve efficiency and effectiveness, to enable organizations to do more with less, to produce higher-quality outputs, and to increase the value-added content of their products and services. Companies can leverage their data to design products that better match customer needs. No more guessing what the customer wants. Through in-store behavior analysis and customer microsegmentation, companies can optimize market segments and know exactly what the customer is buying. In fact, analytics is moving businesses into an era of "shopper marketing"—monitoring and creating an entire shopping experience for customers no matter where they are along their shopping path: at home (online), on the go (through geolocation), and within stores (in-store monitoring).[35]

Data can even be leveraged to improve products as they are used. An example is a mobile phone that has learned the owner's habits and preferences—that holds applications, photos, and data tailored to that particular user's needs. That device will therefore become more valuable with use than a new device that has not become customized.[36]

1.4.3 Creating New Business Opportunities

The information potential of data is opening all kinds of new business opportunities. Consider the possibilities from the mere ability to gather data about how car parts are actually used on the road. This data can be used to improve the design of parts and is turning out to be a big competitive advantage for the firms that can obtain the information. Consider the company Intrix, which collects geolocation information. In 2012, the company ran a trial of analyzing where and when the automatic braking systems (ABS) of a car kicked in.[37] The premise was that frequent triggering of the ABS on a particular stretch of road may imply that conditions there are dangerous, and that drivers should consider alternative routes. With this, Intrix developed the service offering to recommend not only the shortest route, but the safest one as well. This is an entirely novel area of business.

Big data is also helping create entirely new types of businesses, especially those that aggregate and analyze data. Data is the new asset and most organizations are unable to keep up with its rapid growth, scale, and evolution. This is not their core competency. As a result, most non-IT companies are turning to some solutions providers for help. Companies are routinely outsourcing this capability and turning to third parties. This is the rise of the *third-party analytics provider (3PA)*. These are various analytics and IT experts, data brokers, software vendors, and solutions consultants. Similar to third-party logistics providers (3PLs) that orchestrate the movement of physical goods, these companies coordinate and make sense out of large data flows.

1.5 Data-Driven Supply Chains

Few areas of business have been transformed by big data analytics as much as supply chain management. Same-day delivery has become nearly mandatory to modern multichannel retailing.[38] As consumers, we have developed this expectation. We don't think about it unless there is a problem. It may be that the item we ordered online doesn't show up as scheduled or an advertised item is out of stock when we try to purchase it. Achieving a competitive level of global supply chain excellence cannot be accomplished without data-driven, end-to-end operations.

Consider companies such as Tesco. The company gathers huge amounts of customer data from its loyalty program. It then mines this data to inform decisions from promotions to strategic segmentation of customers. Amazon came early to the frontier of data analytics. The online retailer pushed the frontier using customer data to power its recommendation engine "you may also like..." based on a type of predictive modeling technique called collaborative filtering. The company continues its rapid leadership in fulfillment capabilities through data-driven decisions. Walmart was also an early adopter of data-driven supply chains. By making supply-and-demand signals visible between retail stores and suppliers, the company optimizes all its supply chain decisions—from customer fulfillment to inventory tracking (think POS data and RFID sensors) to automatic purchase orders through its supplier portal.

The number of RFID tags sensing inventories across supply chains is in the millions. In fact, the number of RFID tags sold globally is projected to rise from 12 million in 2011 to 209 billion in 2021.[39] Supply chains are increasingly combining data from different systems to coordinate activities across the supply chain end-to-end. Marketing is generating huge volumes of POS data from retail stores that is automatically shared with suppliers for real-time, stock-level monitoring. RFID tags monitor inventory on shelves and in-transit coordinating with current stock levels for automatic order replenishment. Add to this data from computer-aided design, computer-aided engineering, computer-aided manufacturing, collaborative product development management, and digital manufacturing, and connect it across organizational boundaries in an end-to-end supply chain.

Even more value can be unlocked from big data when companies are able to integrate data from other sources. This includes data from retailers that goes well beyond sales. It may be promotion data, such as items, prices, and sales. It also includes launch data, such as specific items to be listed and associated ramp-up and ramp-down plans. It also includes inventory data, such as stock levels per warehouse and sales per store. This data is essential for the supply chain to deliver the items that are needed when they are needed.

Through collaborative supply chain management and planning, companies can mitigate the bullwhip effect and better smooth out flow through the supply chain. Many companies guard customer data as proprietary, but there are many examples of successful data sharing. Walmart is a great example of requiring all suppliers to use its Retail Link platform.[40] The exchange and sharing of data across the extended enterprise has provided transparency and enabled coordinated cross-enterprise efforts.

Big data analytics is the game changer. It has given rise to the *intelligent supply chain*.

2

Transforming Supply Chains

Walmart is a leader among supply chain analytics competitors. The company exemplifies the benefits that can be achieved when applying analytics along the entire supply chain. Yes, the company collects massive amounts of sales and inventory data—more than one million customer transactions every hour.[1] And all that data goes into a single integrated technology platform.[2] Here, managers routinely analyze numerous aspects of their supply chain along all supply chain levers. They also generate *integrated* and *coordinated* data-driven decisions.

Walmart collects massive amounts of information on consumer behavior that it stores in its data warehouse. In fact, Walmart now collects more data about more consumers than anyone in the private sector. The company then applies numerous analytics on this massive database. Who uses this data? Everyone across Walmart's supply chain organization.

Managers at the store level use the system to analyze detailed sales data and use analytics to optimize product assortment. They also examine qualitative factors to tailor assortments to local communities. Marketers at Walmart mine data to ensure that customers have the products they want, when and where they want them. Using analytics, the company has learned that before a hurricane, consumers stock up on food items that don't require cooking or refrigeration. Analytics has uncovered that, in particular, consumers want Kellogg's Pop-Tarts—and not just any Pop-Tarts, but Strawberry Pop-Tarts. Walmart can then work with Kellogg's to rush shipments to stores in preparation for a hurricane. It is this detailed customer tracking that gives Walmart deep insights into customer preferences and buying behavior. Leveraging this knowledge enables Walmart to win important pricing and distribution concessions from its suppliers.

Walmart's analytics and data availability extends to all its suppliers—more than 17,400 suppliers in 80 countries. Each of the suppliers is required to use the company's "Retail Link" system to track the movement of its products—a tool that gives its suppliers a view of demand in stores. Suppliers know in real time when stores should be restocked. They don't have to wait for an order from Walmart. Suppliers can search for information on sales, shipments, purchase orders, invoices, claims, and forecasts. They can run queries on the data warehouse. Suppliers also have access to Walmart's assortment planning system. They can use the system to create store-specific modular layouts based on sales data and store traits.

In short, Walmart's success as the world's largest retailer is at least in part based on the many analytical applications it uses to manage its global supply network. This success is not only because the company has been a pioneer in applying big data analytics, but also because the company uses big data analytics to link its entire supply chain.

2.1 Across the Entire Supply Chain

All supply chains have four levers in common: Buy, Make, Move, and Sell (see Figure 2.1). Leading companies know that focusing on any one lever is not enough. Walmart, for example, uses big data analytics to link the entire chain from Buy to Sell. It is the Sell side that captures and tracks demand through POS data. This information moves efficiently through the supply chain to inform all the other levers. This is where Walmart's Retail Link system kicks in and informs the other levers, such as connecting the Buy side with the Sell side. Information at any one lever—say a shortage on the Buy side, delayed shipments on the Move side, or production stoppage on the Make side—is conveyed to other levers informing them and coordinating action.

Figure 2.1 Supply chain levers

2.1.1 Linking the Levers

Using big data analytics to link the supply chain levers creates tremendous opportunities for a competitive advantage. It is through these analytical tools that companies can develop close relationships with customers based on a deep understanding of their behaviors and needs. They can deliver the targeted advertising and exact promotions that motivate customers to buy and create loyalty. They can charge exactly the price that customers are willing to pay at any moment—and change pricing based on factors in real time. Companies can determine the best use of marketing investments. They can determine the optimal locations for their stores and distribution centers. Most important, they can balance inventory with demand so they are never out of stock or carry excess inventory.

Leading firms are doing exactly this and have already achieved dramatic benefits. CVS, for example, uses analytics to generate targeted coupons at point of sale and views its analytical capability as a large profit center. CVS conducts analytical analysis of data it collects from its ExtraCare customer loyalty system. In fact, 70 million people have used the card over a recent six-month period.[3] The company then shares this information with those suppliers that sponsor promotions to ExtraCare members through its "Partner Portal."[4]

2.2 The Supply Chain System

To effectively link all the supply chain levers, we need to manage the supply chain as a system. There are some rules for doing this. In this section we look at these rules and provide some general best practices for supply chain system design, including optimization, strategy, and network design.

2.2.1 Do Not Suboptimize the System

Analytics applied exclusively to any single supply chain lever will not optimize the supply chain. The reason is that a supply chain is a *system*. And as in any system, all the levers—Buy, Make, Move, and Sell—need to be linked for it to work properly. Otherwise, cutting costs along one lever often results in increased costs in another. Marketing, for example, may use big data analytics to customize product offerings. However, if operations is not able to produce the desired product versions and quantities, or if logistics is not prepared to deliver them, the system results in higher costs. The results are higher inventories, higher setup, or poor customer service. This is the classic example of how companies get into trouble—*suboptimizing the system.*

A supply chain can be viewed as a system of processes that cut across organizations and deliver customer value, rather than as a series of separate organizations and functions. In this case, the focus is not just to manage each process within the organization, but to manage processes across the entire supply chain.

Consider the situation faced by ASICS—a leading maker of athletic footwear, sports apparel, and accessories. In 2008, the North American branch found sales growing 21 percent annually. The company found it difficult to keep ahead of demand. ASICS had only one distribution center in the United States, which had reached capacity. This single distribution center (DC) located in Southaven, Mississippi, was able to handle a maximum of 50,000 units per day. However, the growth in demand resulted in 70,000 units per day being shipped to the DC. This capacity constraint was not only slowing down order fulfillment, but it was also now preventing the company from serving new customers and markets. The DC had become a bottleneck in the supply chain network. The result was customer service slowdowns and high inventory. The company understood that focusing just on the Sell side without matching it with other supply chain levers was not going to work. The supply chain network itself had to be changed if the company was going to support this new level of demand.

ASICS turned to Fortna, Inc., a consulting company. After analyzing the network and current demands, Fortna redesigned the network to make sure that all the levers worked together to support the Sell side. This involved shifting some distribution operations and constructing a second distribution center. The bottom line was that the supply chain network needed to work as a system. Growth on the Sell side was not supported by the Make and Move sides, in this case resulting in high costs and lower service levels.

The lesson here is that the supply chain network needs to be improved as a system. Implementation along only one lever will not optimize the system. Like a tube of toothpaste squeezed in the middle, excess inventories will bulge out in other spots. Big data analytics needs to be implemented so it connects the levers across the supply chain.

2.2.2 Let Strategy Drive

Over the years, countless companies have rushed into implementing the process or technology of the day. This type of "herd mentality" was seen with implementation of lean systems, theory of constraints (TOC), Six Sigma processes, and countless others. But not everyone needs every new technology or the same level of implementation as others do. Strategy needs to drive what—and how much—the company needs.

Supply chain strategy provides the long-range plan for this entire system. Two key elements support this strategy. The first is the *supply chain network design,* which includes the physical structure and business processes included in the system. These are the parts that make up the Buy, Make, Move, and Sell levers. The second is the *information technology (IT),* which enables data sharing, communication, and process synchronization (see Figure 2.2). This is the "home" of big data analytics. IT is the backbone of supply chain management that enables managing processes. Without IT communication and coordination, decision making across the supply chain could not take place. Together, the supply chain network design and IT system design support the supply chain strategy. These elements have to be aligned and work in unison as one system. This includes the implementation and application of big data analytics.

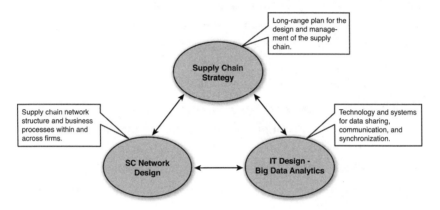

Figure 2.2 Network and IT design support supply chain strategy

2.2.3 Network Design

Individual companies rarely succeed with analytics by "going it alone." Instead, they collaborate with multiple partners in a symbiotic network often called the *analytical ecosystem.*[5] The network may consist of suppliers, channel partners, and distributors as well as external providers of data and analytical services, and software and hardware providers. In fact, there is an increasing growth of analytical outsourcing and the emergence of "third-party analytics" providers.

Companies typically lack the capabilities to do all the analytical work that is required to maintain competitiveness. As a result, they are turning to external entities for help. Most companies need software from external providers and may need external data as well. Recognizing this need, many providers of software and data are now also providing analytical assistance.[6] Their expertise on data sources, platforms, and analytical techniques makes them uniquely qualified to help with analysis. Some provide expertise that cuts across industries. Others target particular industry sectors developing deep, industry-specific skills. In general, using external providers helps companies quickly scale up their analytical expertise. This is often a faster and less-expensive approach compared with trying to do it alone. The trend of outsourcing IT capability will continue to increase as technological expertise continues to evolve and require highly specialized knowledge.

Examples of such outsourcing arrangements abound. Accenture, for example, provides consulting and outsourcing services to a variety of companies. One of its clients is Best Buy.[7] Accenture assists with analytical strategy and analytical applications and manages major components of Best Buy's IT function. Similarly, software firm Teradata worked closely with Hudson's Bay Company to implement a new approach to reducing fraud in the merchandise returns process. Alliance Data works with retailers, such as Limited Brands and Pottery Barn, to establish and manage their loyalty programs. Mu Sigma provides analytical services to Walmart and other retailers.

These types of outsourcing arrangements are almost always necessary as companies do not have the skill set to acquire and keep up with the technology. However, as with any outsourcing arrangement, companies need to focus on their core competencies and, as in the famous words of Peter Drucker, "Do what you do best and outsource the rest."

A well-thought-out analytics outsourcing strategy is a must. This strategy must clearly delineate which analytical capabilities the company wants to build for itself, and which they will outsource to external providers. The strategy must also specify the long-term plan for building capabilities over time, in addition to meeting short-term goals. Lastly, a part of this strategy will be whether to follow the lead of the "supply chain master"—dominant companies in a supply chain—or try to build independent capabilities. This is an important strategic decision that will include issues of market position, size, and long-term growth plans. Consider, for example, that Walmart's suppliers are required to use its Retail Link portal. Although the portal provides reports and some analytics on suppliers' sales by store, promotional activities, and inventory levels, the company maintains control over the data.

2.3 From Sourcing to Sales

Leading companies are using big data analytics along all supply chain levers—from Buy to Sell. Big data applications have improved every aspect of the supply chain (see Figure 2.3). On the Sell side—marketing—applications are rapidly growing to better understand consumer behavior. The Move side—logistics—has used big data analytics applications for routing and vehicle scheduling for years. The Make side—operations—is increasingly using applications that include inventory and capacity optimization. Lastly, the Buy side—purchasing—is increasingly seeing applications that better evaluate supplier performance and inform supplier negotiations.

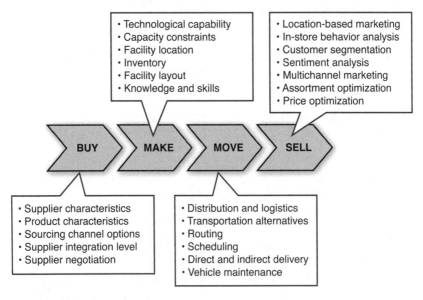

Figure 2.3 Big data enables supply chain levers

Consider Intel. The company lists numerous advantages of big data analytics.[8] These include providing data visualization that gives the organization supply chain visibility. Using big data analytics, Intel is able to examine the full spectrum of supply chain network decisions, such as determining promised order fulfillment lead times, inventory needs, and transportation and storage costs. Modeling or simulations are used to take the large data pools and turn them into meaningful information. For Intel, big data analytics goes beyond tactical supply chain decisions. It helps the company look at broad issues, such as supply chain compliance, social responsibility, and sustainability—from identifying supplier problems to managing its carbon footprint.

2.3.1 Buy (or Source)

The Buy lever is essentially sourcing or purchasing. It is the business function responsible for all activities and processes required to purchase goods and services from suppliers. Although marketing has the primary responsibility for the customer side of the organization, it is sourcing that has the primary responsibility for the supply side.

Sourcing also has a critical financial impact on the organization. Companies spend large sums of money on sourcing goods from their suppliers, and the savings that can result from proper management of this function can have a huge impact on the organization. Consider that in most manufacturing organizations, the sourcing function represents the largest single category of spend for the company, ranging from 50 to 90 percent of revenue. In fact, almost 80 percent of the cost of an automobile is purchased cost, where the manufacturing facility merely assembles purchased items.[9] Applying big data analytics along this lever can yield great savings.

Big data analytics is critical in sourcing. A number of leading companies report using big data to optimize sourcing channel options and integrate suppliers into data systems. Some mention using big data to help identify supplier characteristics and to help inform their supplier negotiation. Some companies are using it to analyze vendor data and come up with a recommendation of which vendor to procure products from for various criteria, such as cost or risk.

Amazon, for example, uses analytics to determine the optimal sourcing strategy as well as manage all the logistics to get a product from manufacturer to customer. This includes using analytics to determine the right mix of joint replenishment, coordinated replenishment, and single sourcing. In fact, Amazon applies advanced optimization and supply chain management methodologies and techniques across its fulfillment, capacity expansion, inventory management, procurement, and logistics functions.[10]

On the Buy side, big data can be used to inform supplier negotiations. Big data analytics can be used to analyze customer preferences and buying behaviors. This, in turn, can be used to inform negotiations with suppliers. Companies can use information on prices and transactions to negotiate concessions on key products.

2.3.2 Make

Make is the operations function of the organization. It is responsible for creating goods and services. Make—or operations—serves a transformation role in the organization by converting a company's inputs into finished products. These inputs include materials, technology, information, human resources—such as workers, staff, and managers—and facilities and processes—such as buildings and equipment. Outputs are the goods and services a company produces—from health-care services to toys, cars, and clothes.

Analytics is used extensively along this lever to improve all aspects of the Make process. A large number of companies report using big data analytics for inventory management, optimization of stock levels, maintenance optimization, and deciding upon where new facilities should be located. Some companies are considering using big data analytics in workforce productivity evaluation as well as in the study of capacity constraints.

Productivity and quality have huge applications. With the help of big data analytics, companies can now run daily analyses of performance. These statistics can be aggregated and reported by store sales, stock keeping unit (SKU) sales, and sales per employee. These systems are now moving ever closer to real time where they can alert companies of problems, such as changes in productivity or a breach in quality. Companies can look at labor data for accuracy and quality performance. Although companies have been using technology for these metrics for many years, the difference is the scale. The trend is toward much higher frequency, immediacy, and granular reporting of these statistics. This allows managers to make more targeted adjustments in their operations in a much more timely manner.

Labor optimization is another operational lever that utilizes big data analytics. These technologies can enable reducing costs while maintaining service levels by optimizing labor, automating and tracking attendance, and improving labor scheduling. For example, retailers can analyze cashier performance such as transactions per hour. Managers at a call center can analyze quality of customer service based on customer complaints, satisfaction surveys, or the percentage of customer issues solved with a single call. Analytics can also create more accurate predictions of staffing needs, by matching good forecasting with labor optimization. This can be especially beneficial during peak demand periods.

2.3.3 Move

Move is about logistics. Logistics is the business function responsible for transporting and delivering products to the right place at the right time throughout the supply chain. Logistics provides movement and storage of product inventories throughout the chain. The decisions involved range from measuring inventories to planning and coordinating material flows to arranging distribution routes and shipping. Big data applications have been used along this lever for optimizing inventory, replenishment, identifying optimal distribution center locations, and minimizing transportation costs. A number of the companies report using big data analytics for vehicle maintenance, routing and scheduling, and selection of transportation alternatives.

Analytics applications along this lever are some of the oldest. Supply chain analytics has involved optimization of locations, inventory levels, and supply routes for many years, having been studied under the term *operations research*. Just as analytics can be used to segment customers, companies are segmenting transportation routes, modes of transportation, and transit factors for different types of products.

A big area for analytics is in transportation and routing. For example, leading companies are optimizing transportation by using GPS-enabled big data telematics and route optimization. Transport analytics can improve productivity by optimizing fuel efficiency, preventive maintenance, driver behavior, and vehicle routing. Tracking of weather and other disruptive events can continually optimize routes. UPS, for example, started gathering data more than 20 years ago. Today, the company uses a "data-drenched" tool called ORION (On-Road Integrated Optimization and Navigation) to help drivers find the most efficient path through their delivery areas.[11]

Another area is inventory management. Although inventory management is an issue at all locations—from stores, to warehouses, to manufacturing facilities—it is especially problematic in transit. RFID technology has been especially useful in tracking inventory in motion, identifying location and quantities, and preventing security breaches. The technology can also monitor temperature en route, ensuring food safety, monitoring transit duration—which is an especially important issue for perishable items—and informing decision makers if there is a problem.

2.3.4 Sell

The Sell lever of the supply chain is marketing. Marketing is the function responsible for linking the organization to its customers, identifying what customers need and want, generating demand for current and new products, and identifying market opportunities. For an organization and its supply chain to be competitive, they must be better than competitors at meeting customer needs. This is the responsibility of the Sell lever. This has driven the development of big data applications to capture customer demand, enable microsegmentation, and predict consumer behavior.

Microsegmentation is a highly important application of big data analytics. Although creating segments is familiar to marketing, big data analytics has enabled tremendous innovation in recent years. The coupling of big data with increasing sophistication in analytic tools has allowed microsegmentation at increasingly granular levels. Companies can now gather and track data on the behavior of individual customers. They can then combine these with traditional market research tools for greater insight. As the collected data becomes more granular and as it is increasingly tracked in real time, companies can readjust their strategies to customer changes as they occur. Neiman Marcus is a retailer that combines behavioral segmentation with its multitier membership reward program.[12] It uses a sophisticated analytics program to identify, customize, and incentivize purchases. Targeting the most affluent customers has led to a significant increase in higher-margin purchases from the company's higher-margin customers.

Another application on the Sell side is in price optimization. Increasing granularity of data on pricing and sales coupled with sophisticated analytics has taken pricing optimization to a new level. A variety of data sources can be used to evaluate and inform pricing decisions in near real time. Consider Marriott International. The company uses a sophisticated analytics system to establish optimal prices for guest rooms through its revenue management program.[13] Numerous factors are included and adjusted in real time—from type of customer to weather.

Applications designed to understand consumer behavior are increasingly becoming more sophisticated. In fact, Sell is the supply chain lever that has seen a tremendous growth in big data applications. Companies using big data analytics along this lever are going well beyond customer and market segmentation. They are using it for every aspect of understanding and tracking consumer behavior—from location-based marketing and sentiment analysis, to in-store behavior analysis. Some retailers are also using it for merchandising, particularly price and assortment optimization. This intelligence is used to drive entire supply chains, as demonstrated by Walmart, which links this data through the chain to coordinate all activities at stores, warehouses, in transit, and with its suppliers.

2.4 Coordinated and Integrated

Big data analytics applications enhance performance across all supply chain levers. Each lever offers capabilities that turn information into intelligence and improve functionality. Although these applications offer unprecedented insights, these efforts should not be done in a fragmented fashion. To achieve a competitive advantage, these efforts need to be part of a *coordinated and integrated overarching strategy* championed by top leadership and pushed down to decision makers at all levels of the organization and across the supply chain. The decisions need to align across the enterprise and the supply chain. They also need to align vertically from the very top of the organization to the very bottom. Leaders need to create a culture that supports evidence and fact-based decision making. All these efforts need to be aligned and integrated. An analytics supply chain network is created when this vision is propagated through the supply chain. Data and analytics optimize decisions, technology automates the process, and decisions are communicated and coordinated from Buy to Sell. This creates the intelligent supply chain.

Coca-Cola provides an example of an aligned and analytics-driven, end-to-end supply chain. The cola giant has an algorithm to engineer the taste of its orange juice. A computer model directs everything from picking schedules to the blend of ingredients needed to maintain a consistent taste. Coke has spent $114 million to expand its technology-driven U.S. juice-bottling plant, which it claims to be the world's largest. It is here the company uses a methodology it calls Black Box. This is not a secret recipe. It is an algorithm. Black Box includes detailed data from more than 600 flavors that make up the "taste" of what customers perceive as an orange. That data is matched to a profile detailing acidity, sweetness, and other attributes of each batch of raw juice. The algorithm then tells Coke how to blend batches to replicate a certain taste and consistency, including the amount of pulp to include. The algorithm is then tied to satellite imaging of fruit groves, ensuring fruit is picked at the optimal time for Coke's bottling plants. Every aspect of the supply chain is optimized. The algorithm also includes external factors, such as crop yield, current prices, and weather patterns. The mathematical model can quickly create new plans—in a matter of five or ten minutes—in light of any new information. Everything has been standardized and optimized. According to Jim Horrisberger, director of procurement, "You take Mother Nature and standardize it."[14]

2.5 The Intelligent Supply Chain

Big data analytics applied across all supply chain levers creates the intelligent supply chain. Let's look at some of the forward-looking supply chain applications of IT, which go beyond traditional marketing uses to track and monitor inventory movement, creating an information feedback loop between sales information and supply chain management.

2.5.1 Not Just for Marketing

Most of us have become accustomed to the focus of companies collecting sales data to analyze customer behavior. Think of Amazon and Netflix offering "you might also like" lists, which are derived from tracking buying behavior and reviews. The popular press and news media have put great emphasis on the power of big data analytics to understand customer preferences. Consider the recent publicity made by retailers strategically placing facial recognition technology in eyes of mannequins to capture customer movements. The Eye-See is a mannequin with a camera that has facial recognition software embedded in one eye.[15] The information gathered is much richer than what can be obtained through security cameras. This technology gathers knowledge on customer profiles—such as age, gender, and ethnicity—and then couples that with shopping behavior. This enables retailers to better manage their merchandise and enhance business profitability.

What most of us are not aware of, however, is the huge move by companies to leverage the big data from all supply chain functions in order to enhance operational efficiencies and reduce costs. Data is collected and communicated along all supply chain levers, communicating information, signaling events, and enabling coordination in real time. This is the intelligent supply chain. Big data is streaming from all supply chain levers and operations. Applying analytics to the data optimizes the system and processes, automates decision making, and increases overall efficiencies.

An excellent example of creating an intelligent supply chain is Amazon. Amazon has built a new supply chain model that links most supply chain decisions and optimizes them mathematically. Amazon even hired a new team of supply chain analysts and operations researchers in order to create the new model.[16] It is a *nonstationary stochastic model*—or in lay terms, *complicated.* The model is used to determine fulfillment, sourcing, capacity, and inventory decisions. Amazon developed new algorithms for joint and coordinated replenishment. It also implemented a new forecasting approach at the SKU level. The model is based on historical demand and considers event history and future plans. It makes forecasts for each fulfillment center, inventory planning, procurement cycles, and purchase orders. In essence, the model connects the supply chain. Just consider the amount of data that would be involved in forecasting at such a granular level for so many purposes and at so many levels.

2.5.2 Game-Changing Technologies

Two technologies have converged to enable the intelligent supply chain. These are POS data and RFID technologies. Although both have been around for decades, the new generation of these technologies is unprecedented in its capabilities. Together, these technologies create data streams that can be leveraged to better understand demand and manage inventory, supplies, and resources. Together, they can be used to link the levers of the supply chain.

2.5.2a POS Data

POS (point-of-sale) transactional data first appeared in the 1970s and was primarily obtained from the use of bar codes. Today's POS technology, however, captures data from stores and other selling channels in real time and in ever-more sophisticated ways. This data can now be more accurately and granularly captured, analyzed, and used.

The real-time POS data captured at store locations holds an infinite amount of information that can be used to manage the supply chain. It also holds information on quantity, price, discounts, and coupons being used. In addition, this information contains location data. The information helps identify what products are moving and how fast they are moving. It can also be used to conduct geographic analysis and customization.

Now imagine tying this information to loyalty cards. Coupling such data can enable high levels of segmentation, creating an endless array of customer profiles. All this can help manage demand better by evaluating and optimally deploying inventory. This can prevent stock-outs at the most profitable stores—where it matters most. This data can also be used to continuously evaluate prices and other incentives to enhance overall profitability. It can also help manage seasonal ramp-downs and reduce the need for clearance events.

2.5.2b RFID

RFID tags have been used for many decades and were originally invented for military purposes. Today's technology, however, is much less expensive and better than that of just a few years ago. RFID tags are used to identify and track items—and have found numerous applications across all industry sectors. For example, RFID tags attached to an automobile during production can be used to track its progress through the assembly line. Pharmaceuticals can be tracked through distribution and warehousing. The RFID tags can also help in restocking inventory, making sure the items are put in the correct place.[17]

One advantage of reading RFID tags is that line of sight is not required to read them. This enables inventory management to be performed in a highly efficient manner. For example, RFID tags can be used to read pallets in a warehouse. The items can be identified and counted, and their location can be noted. This can be determined no matter where the tag is placed on the pallet. RFID tags are slowly becoming embedded in all parts of the supply chain and are the driving factor behind the supply chain management big data wave.

2.5.3 Dual Power: POS and RFID

Now, couple the consumer POS data obtained in real time with intelligent, automated inventory ordering and replenishment systems in the store through RFID tags. The tags can respond to the real-time consumption in the store to ensure almost zero stock-outs. This has enabled the retail shelves to assume the role of the inventory manager. The tags can send alerts for restocking the shelves when item quantities fall below the set threshold. They can also order store replenishments when the store supplies fall below inventory thresholds. Both item-level RFID tags and POS technologies have become ubiquitous and represent the state of the art.

Now let's tie this to the rest of the supply chain. Using RFID tags, warehouses can also become intelligent. They can continuously monitor stocks of inventory and send replenishment order requests as needed. Purchase orders can also be automated and the same tags can track the inventory-in-transit in real time. Data and analyses can also be made available across the extended supply chain to manufacturers and distributors.

Supply chain analytics should also be jointly determined and analyzed by supply chain partners. For example, companies can jointly engage in efforts such as Collaborative Planning, Forecasting, and Replenishment (CPFR). Together, supply chain partners can establish key supply chain metrics and engage in iterative planning and forecasting supported by analytics applications. The electronics retailer Best Buy is a good example of this. The company uses CPFR to exchange reports and analyses with its major suppliers. The company integrates all data on forecasts and inventories in a joint database with its trading partners.

But the supply chain can be even more intelligent. How?

As analytics capability grows, companies will be able to gather more intelligence from the information and create superior supply chains. Applying analytics can determine product quantities, identify optimal distribution center locations, optimize inventories and replenishments, and minimize transportation costs. The ability to collect, analyze, and use real time demand and inventory data is becoming increasingly more accessible. As a result, big data analytics will enable optimizing supply chain operations and provide competitive advantage to those who can use the technology to support their business strategy and enhance their competitive priorities. One such advantage will come through more granular segmentation across all levers. Just as companies segment customers and markets, they are now segmenting supply chains and aligning products to match different supply chain processes. This segmentation can be based on numerous factors, such as product cost, shipping cost, perishability, innovation rate, or any number of factors. Similarly, companies are segmenting their suppliers, based on responsiveness, quality, cost, or risk. Lastly, big data analytics can be used to match these segments—supply and demand—and design optimal supply chain networks.

The German department store retailer Metro Group is using this type of intelligence.[18] The company is testing its use of RFID data to detect the movement of goods within stores. The information can track patterns of movements, monitor inventory levels, and create out-of-stock alerts. The information can also identify a product and offer recommendations for related products, all done in real time and informing the customer whether the related products are in stock.

Similarly, office products retailer OfficeMax has put in place a new analytics system for supply chain management.[19] The system is designed to optimize inventory, transportation cost, and warehouse investment, while maintaining the highest in-stock availability. The model uses data to analyze in-store product movements, which drive restocking policies as well as rearranging assortments to maximize sales.

2.5.4 Moving Forward

Leading-edge companies are recognizing that the use of technology and big data analytics is needed along the entire supply chain. Although most are using big data analytics on the demand side of the supply chain, more will move to using it along the supply side and to better coordinate supply with demand. More companies are using some type of big data software and analysis to drive their entire supply chains. Almost all supply chain organizations recognize this to be a competitive necessity. Big data is being used along all supply chain levers, and successful applications require coordinated decisions across organizations and along the entire supply chain.

There are three major challenges companies face. First, companies need to ensure that they don't follow the hype but use the technology to meet their own competitive priorities in a cost-effective manner. Second, companies need to develop organizational processes to turn the huge amounts of data into business intelligence. Third, companies need to overcome common barriers associated with moving forward. The next chapter looks at what these are and how they can impede implementation. Then subsequent chapters look at a roadmap to overcome these barriers and a plan for creating the needed transformation to move forward.

3

Barriers to Implementation

Even small companies are doing it.

Adoption of big data analytics has not been restricted to large firms—firms such as Amazon, UPS, IBM, and Walmart. The Wine House is a midsized retailer of wine. It has no sophisticated analytics systems. Still, the company decided that it needed to analyze the movement of its inventory.[1] Using simple analytics, the company identified 1,000 cases of wine that hadn't sold for more than a year. With further analysis, the firm has learned which price points sell best and which suppliers' products are most profitable. Similarly, the English fashion retailer Jaeger is using data mining to identify losses resulting from employee theft and process-related errors. After only three months of analysis, Jaeger determined that its savings were significantly more than predicted before implementation. These small companies are not using the most advanced and expensive analytics and IT systems. Nevertheless, they are gaining benefits from big data analytics.

3.1 Why Isn't Everyone Doing It?

Big data software and analysis will be the most important supply chain technology for forecasting, demand planning, and supply chain management in the years to come. Through analysis of huge quantities of data, big data analytics provides a competitive advantage by providing unparalleled insights. The challenge for companies will be staying ahead of the technology in a cost-effective manner and developing organizational processes to effectively utilize the huge amounts of data and absorb the information into their organizational decision-making processes—and turn it into intelligence.

A recent survey of 214 senior executives and managers finds the majority (78 percent) to believe big data analytics is a technological priority for the future.[2] Sixty-eight percent expected to be making some degree of supply chain technological software investment in the coming year. However, 83 percent expressed substantial concerns regarding costs of the technology and the *adoption* decision for their individual needs, considering costs versus capabilities needed. This was especially true of smaller and medium-sized firms. Further, they expressed concern regarding *adapting* the technology to their organizational processes.

The following sections discuss the specific challenges identified.

3.1.1 Data Versus Information

Big data enables mining of huge quantities of data. The trick is converting this massive data into usable information the organization can absorb and effectively use to make decisions. There are two critical elements for success. First, the organizations must have a good process in place before technology adoption. Technology overlaid on top of poor processes just solidifies poor performance. Also, a rudimentary knowledge of techniques and methodologies continues to be essential. Consider decision areas such as forecasting. Relying on analytics does not mean abdicating an understanding of how techniques work and relying on the output blindly. Leading companies understand that they cannot just invest in these technologies without the adequate organizational processes in place.

Second, organizations need to have an infrastructure that can absorb the information generated by analytics systems into organizational decision-making processes. An infrastructure of processes and workflows enables information to actually be at the place where and when it is actually needed. This requires creating a learning organization where processes are in place to absorb the new information.

3.1.2 Customer Service Driver

One of the primary reasons for investing in analytics software is to improve operational performance. This includes improving forecast accuracy, reducing demand variability, and improving supply chain visibility. However, another high-ranking reason is to improve customer service. In fact, a large percentage of companies today are placing greater importance on enhancing the customer experience, as a driver of growth, rather than merely cutting costs. Leading manufacturers are expanding their product offerings to include a service component—called *servitization*.[3] Adding services to the product offering can enhance competitiveness, but it also introduces many supply chain complexities. Analytics helps identify key elements of the product-service bundle customers consider most important. This then feeds back up the supply chain to product design and co-creation with suppliers. The challenge is to create these linkages to enable end-to-end functionality that is driven by customer service.

3.1.3 Dismantling Information Silos

Past investments in sophisticated technologies—such as customer relationship management (CRM) or supplier management software—have created information silos at many organizations. Using insights from big data, companies can better integrate information residing in disparate organizational areas, such as purchasing, operations, and distribution. In fact, one of the biggest challenges for companies is to dismantle information silos and to connect these islands of information. Effective performance will come from unifying these into a single database.

3.1.4 Sales and Operations Planning

Best-in-class companies are breaking down the silos with big data sharing across the organization, improved collaboration, and use of unifying metrics. Unifying databases from CRM, supplier management, and operations enables better understanding of end-to-end processes. Further, incorporating pricing, inventory, and risk as decision factors within the Sales & Operations Planning (S&OP) and Sales & Inventory Operations Planning (S&IOP) process provides greater understanding. S&OP—discussed in detail in Chapter 9, "Making It Work"—is an excellent process to bring big data analytics into the organization and cut across silos. By its nature, S&OP is a data-driven, cross-functional process. As such, it provides excellent ground for big data analytics implementation.

Companies are increasingly committing resources to new technologies, investing in technology upgrades, and hiring analytical capabilities, such as decision scientists. Leading-edge companies are also redesigning internal processes, such as breaking down silos and improving collaboration, in order to utilize these new information capabilities to their fullest extent. Also, these companies are ensuring organizational processes are in place that can use this new information for improved decision making. There is no doubt that the technology and analytics trend will continue to provide a competitive advantage to leading-edge firms.[4]

For those still struggling with implementation, what are the barriers and how do we overcome them?

3.2 The Barriers

Big data analytics has demonstrated its potential as seen by the applications and benefits achieved by leading-edge firms. However, many barriers to implementation exist. These barriers fall into three general categories: technology, people, and processes. These barriers work together and must be overcome as such.

Companies sometimes make the mistake of investing in one of the categories—for example, technology—with the exclusion of the others (see Figure 3.1). Consider, for example, that IT investment can give workers better and faster tools to do their jobs. However, without investing in people—the human capital and organizational change—there will be no substantial improvement in productivity.

The reason is that investments have to be made both in IT and in managerial processes that complement the IT investment. They all need to be addressed and they need to be able to work together. This is an important consideration. Companies need to pursue their big data analytics investments as part of a large *overarching strategy*, planning their own needs and following their own roadmap, rather than being pushed by the hype.

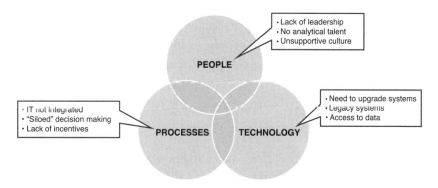

Figure 3.1 The barriers—a mix of people, processes, and technology

3.2.1 Technology

Many businesses have existing technology that might need modification or substitution to capture the benefits of big data analytics. The next section covers some of the issues associated with system upgrades and legacy systems.

3.2.1a Need to Upgrade Systems

There is no question that most organizations will need to deploy new technologies and techniques as they become analytics-driven organizations. Most will require additional investment in IT hardware, software, applications, data, and services. The level of investment will vary considerably depending on a company's current state of IT capability and maturity. Companies will also have to deal with legacy systems and incompatible standards and formats, which often prevent the integration of data and enable more sophisticated analytics. Most companies will need applications of new techniques, such as new types of analyses.

The range of technology applications will differ depending on the maturity of the institution relative to its analytics evolution. Chapter 8, "The Roadmap," looks at the maturity curve and shows that companies cannot just jump into big data analytics and make large investments. Rather they need to move along a maturity curve, following a process, and evolving gradually. Companies need to proceed with caution and avoid the hype, rather than jumping in. Companies should begin with an assessment of their needs and identify gaps in the technology their enterprise has available.

3.2.1b Legacy Systems

A common technological obstacle for many companies is the existence of legacy IT systems. Many of these systems were installed decades ago. Often, new systems were added over time to solve an immediate problem but they didn't necessarily communicate with older systems in place. For many organizations, these additions have resulted in the creation of multiple silos. Each of these silos generates information, but the silos don't communicate with one another, and often they don't even have compatible standards or formats.

As a result, companies are unable to take advantage of sophisticated analytics, new computing power, or the full option of creating end-to-end solutions. The creation of dashboards, and taking advantage of data visualization is impossible as aggregated data cannot be attained. The problem is that these independent systems cannot be readily integrated and analyzed. Attempts to upgrade them can often be costly and ultimately not provide satisfactory results. Companies with such legacy systems need to consider the option of purchasing completely new systems.

Even deploying new IT-enabled systems can present risks and challenges. Sometimes small unanticipated setbacks in technology can create a negative cycle of adoption. This was the case with the application of RFID systems just a few years ago. In the early years, RFID held huge promise in tracking and managing inventories. However, initial applications found the reliability of RFID readers far less accurate than originally expected. This resulted in the need for manual inputs to correct reader errors. This destroyed the productivity gains originally expected. Adoption of RFID slowed and the cost per RFID tag did not drop as predicted. The issue is that technology adoption entails risks and opportunities that need to be carefully considered. Consider that current legacy IT systems of most companies were state-of-the-art technology some years ago.

3.2.1c Access to Data

Indeed, most supply chains are data-creating machines. However, many organizations and their decision makers still don't have access to data that they need. Data is being created but much of it is not in usable form. It has not been prepared, cleaned or "scrubbed," or made available in a form that is accessible to users. In fact, usable data has quickly become a key competitive asset.

Organizations should first conduct an inventory of their own proprietary data. Often companies are surprised as to the data they actually have. They can also systematically catalog other data to which they could potentially gain access. There is a great deal of publicly available data on government, public, and corporate Web sites. Preparing and inventorying such data can be hired out to a third party. Outsourcing this process can quickly create usable data for the organization. In this case, however, companies need to know what they need and what to ask for—which is not always an easy task.

Another source of data is supply chain partners. Sharing data is in the interest of all supply chain members. Companies may just need to come up with a good value proposition or economic incentive to get partners to share. Consider that Walmart shares data with its suppliers through Retail Link, knowing that it is beneficial for everyone.

Lastly, data can be purchased. There are many newly emerged businesses that serve as data aggregators or data brokers.

3.2.2 People

One of the biggest barriers in implementing big data analytics can be lack of leadership. Organizational leaders often lack the understanding of the value big data can bring to the organization. Remember that many organizations still view IT as playing only a supporting organizational role. Even when data is used for operational decisions, leaders may not understand its full capability, such as being a source of competitive advantage. Moving an organization to become an analytically driven competitor takes leadership of transformational proportions. It requires vision, understanding of the full capability of big data, and understanding of how to lead change. Leading such organizational change is discussed in more detail in Chapter 10, "Leading Organizational Change."

A second barrier is simply lack of analytical talent. Organizations need to hire more individuals with deep analytical capabilities. There is much evidence that a serious shortage of analytical talent exists.[5] Companies need to have a strategy in place to go after the best individuals and work to keep the talent that they have in place. Further, organizations need to invest in training for the employees they currently have in place. The training could be extensive or it could be a simple class in statistics. One option is for organizations to bring in university faculty to conduct short, targeted, and customized educational classes. These are often less expensive and much more effective.

There is one important point here. Not everyone needs to be an expert on analytics. The key is that an organizational culture is present that supports data-driven decisions and that even employees who are on "the front lines" understand the capabilities analytics can bring.

A third barrier is an unsupportive culture. Culture is an elusive characteristic of the work environment. It is the shared values, beliefs, and assumptions of organizational members. A culture that supports decisions based on hunches and guesses needs to be changed for an organization to move to one that is based on analytics. That rests on the shoulders of leadership.

3.2.3 Processes

Many organizations rush into implementing new IT systems and applications in an effort to move into the big data age. They will also hire analytically savvy talent with deep knowledge. However, this will still not lead to success unless the processes that hold these pieces together have been changed accordingly.

One large problem for organizations is simply not having IT integrated into workflow. What good is superior analytical capability if it is not integrated into decision making when it is needed? In many organizations, IT is still a backroom operation that plays a support role. For organizations to achieve excellence with big data analytics, this mind-set must be overcome. Organizations need to restructure workflow so that IT is integrated into decision making in a meaningful way. This will come from joint efforts between functional managers and IT staff in integrating capabilities and needs.

Further, IT needs to be positioned as the engine for growth and should be included in many organizational initiatives. The reason is that many strategic initiatives and new business opportunities are driven by big data, in addition to improvements in efficiency and productivity. Leadership needs to see IT as a key source of a competitive advantage, they need to include IT in discussions, and they need to actively collaborate with IT.

Another problem organizations consistently face is "siloed" decision making. Analytics needs to be a cross-enterprise effort and it needs to connect and coordinate the organization and supply chain. That is a key element of the success of analytics leading firms, such as Tesco, Walmart, UPS, and others. An excellent place for integration of big data analytics is in S&OP processes as well as CPFR. These processes are by definition cross-functional and cross-enterprise and rely on analytics as inputs.

Yet another process barrier is simply lack of incentives. For an organization to become analytics driven requires transformational change. Change, however, can be threatening to employees and the current way of life. As you'll learn in Chapter 10, both intrinsic and extrinsic incentives are needed to motivate people to embrace this change. Unfortunately, many organizations assume that just implementing the IT systems will be sufficient. Ensuring that everyone is on board and that the processes support the new direction is critical.

3.2.4 Analysis Paralysis

The big data revolution offers organizations the ability to capitalize on the massive quantities of data generated by global supply chain activities. Using correlations, forecasting, and predictive analytics over vast quantities of digital information, a few global companies, such as Amazon and UPS, have gained mastery over their supply chains. Big data has enabled greater visibility into inventory levels, order fulfillment rates, delivery of materials and products, and other insights critical to efficient supply chain management. For instance, real-time RFID and SKU reports from transportation vehicles can allow organizations to more precisely time inbound shipments and improve manufacturing productivity. And while companies can use predictive data analytics to match supply with demand, they can also leverage this new competency in sales and operations planning processes to optimize a company's sales channel strategies. Beyond gains in supply chain efficiency and improvements in sales and operations planning, predictive data analytics can inform supply chain strategy and competitive priorities, enable the creation of new business ventures, and create an unparalleled competitive advantage.[6]

Despite these powerful strategic advantages, many companies have yet to leverage big data and predictive analytics to transform their supply chain operations. Many are overwhelmed with the amount of data but are unsure how to use it to drive their supply chains. Some are simply unsure how to even begin. Others are engaging in fragmented utilization or implementation rather than a systematic and coordinated effort. The results are isolated benefits, lack of insight and competitiveness, and supply chains plagued with inefficiencies and cost overruns.

Based on interviews with supply chain management leaders, we identified four hurdles that prevent companies from taking advantage of the big data revolution.[7] These are:

- **A needle in a haystack**—Using analytics randomly in search for causation and relationships with the hope that something will eventually turn up.

- **Islands of excellence**—Some are using analytics to optimize subprocesses; these function efficiently but have little bearing on optimizing the system.

- **Measurement minutiae**—The hallmark of measurement minutiae is trying to measure everything. Few companies have the diligence to actively manage all of the metrics they have created and most companies don't know which ones to focus on.

- **Analysis paralysis**—Many companies are overwhelmed with the rapid change of technological capability; they understand they must do something but are in a state of paralysis.

3.2.5 Drowning in Data

Most companies are drowning in data—from POS systems, Web sites, internal transaction processes, and social media. For most, it is difficult to digest all of the data, technology, and analytics that are available to them. They cannot absorb the new analytical technologies being made available to them. They do not have the necessary systems in place. Many do not have leadership that understands how to lead the organization to become an analytics-driven company—an effort requiring transformational change. Many leaders do not know how to leverage the data capability they currently have. Also, many organizations are not familiar with how to use analytical capabilities for decision making. They don't have the skilled talent to help them improve key decisions. Adding to the challenge is that this is taking place at a time when pressures on global supply chains make it difficult to survive, much less invest to thrive, without the advantage of big data analytics.

Supply chain management in particular is one area that is benefiting from big data. The reason is that the amount and types of data that can be used to manage an end-to-end supply chain have exponentially grown over the past years and are growing even faster. Supply chain–specific data, such as orders, shipments, and inventory levels from retailers, manufacturers, suppliers, and transporters, all provide a rich set of demand signals about products across the value chain. Together, they link supply and demand, from raw materials to final consumption.

Adding to the richness of data are external data sources that improve supply chain performance. An example includes the tracking of weather systems, such as through the weather satellite NOAA. Adding this data, on top of the other sources of information, can design supply chains that can respond to avoid hazardous conditions, areas where stores are closed, or areas where demand spikes for products made for use in inclement weather. The emergence of social media offers yet a new source of potential demand signals. Consider the ability of Google to track the spread of flu through its search engine.[8] Knowledge of such outbreaks can be used to determine which areas may need more chicken soup or cough drops. Add on top of that details about road construction, bridge closings, and weather, and companies can be optimizing routing of shipments in real time so that products continue to reach store shelves on time.

The key issue here is that there is no lack of data in the supply chain. What is missing is the leadership to transform technology, people, and processes—the structures of the systems—to convert the masses of raw data into meaningful and actionable information that can drive the organization and its supply chain.

3.2.6 Misaligned Metrics

Organizations typically employ 10–20 key performance indicators (KPIs) to assess their performance. Many companies, however, add more than that. With analytics, many are adding even more metrics, making it difficult to even know which ones to focus on. Although organizations utilize many metrics in evaluating their supply chain management activities, it is important that they concentrate on those that really measure important things. These are things that make a real difference to the strategy and competitive priorities of the organization, to customer service and firm profitability. These are the factors that everyone involved in supply chain management should know and emphasize as they develop and implement strategies and carry on their day-to-day activities.

The notion that you can't manage what you don't measure is important in all facets of business, especially in supply chain management. The various levers of supply chain management must be measured in order to determine whether systems and processes are being implemented effectively and efficiently Thus, organizations tend to measure "lots of things" in an attempt to plan, implement, and control all the activities.

Successful metrics provide answers to questions such as: "How is our supply chain performing?" and "Where do we need to improve performance?" They also need to include both financial and nonfinancial measures. Not only can analytics help in identifying key metrics of performance, but they can also show how these metrics affect the bottom line.[9]

Organizations respond to what is being measured. Successful implementation of analytics depends on coordination and aligned metrics help in implementation. Metrics need to be process focused, not functionally focused. Organizations need to *align* key metrics to organizational and supply chain management strategies. Consequences of misaligned metrics are many. Most organizations see these. They include the following:

- Purchasing management is rewarded for achieving low raw materials costs.

  The result: Large quantities and higher inventory costs

- Production management is rewarded for achieving the lowest possible per-unit production costs.

> **The result:** Large volume of product produced and high inventory holding cost

- Marketing is rewarded for achieving the highest possible number of sales.

  The result: High production costs or stock-outs

- Logistics management is rewarded for shipping products by full truckload or by railcar to obtain the lowest possible freight rates.

  The result: Higher inventories at manufacturers, distributors, and retailers

- Purchasing management is incentivized to make small and uncoordinated frequent orders in an attempt to reduce inventories and maintain cash flow.

  The result: Inventory carrying costs pushed upstream.

3.3 Breaking Ahead of the Pack

This section explains how the various supply levers can employ big data analytics to create competitive advantages and firm-wide synergies.

3.3.1 Together

Analytics applications that can deliver a competitive advantage appear along every supply chain lever, including Buy, Make, Move, and Sell. Applications range from targeted, location-based marketing on the demand side to refining sourcing channel options and optimizing supply chain inventories on the supply side. What makes this work, however, is coordination and integration. Remember that analytics-driven companies—Amazon, Dell, Walmart, UPS—do all these things in a *coordinated way* as part of an *overarching strategy* in an organization with a data-driven *culture*. This is championed by top leadership with inclusion and support from decision makers at every level. This is the key that cannot be overemphasized.

3.3.2 *The Roadmap*

To break ahead of the pack, companies need a clear roadmap. They need direction to push beyond the barriers and implement big data analytics across their supply chains in a meaningful and cost-effective manner. This book provides a systematic framework for companies on *how* to implement big data analytics across the supply chain to turn information into intelligence and achieve a competitive advantage. Best practices demonstrate that big data driven supply chains achieve a competitive advantage when applied following a systematic and overarching process. This book provides the roadmap to achieve this. We show companies how to focus their big data strategy, use big data at every organizational level and in a coordinated manner, and make the right choices to help the organizations drive competitive global supply chains.

Part II

Impact on Supply Chain Levers

4

Impact on "Sell"

$50 billion. That is the latest estimate on how much money marketers are putting into understanding their customers.[1] So how can big data analytics help? Consider the case of Target. Target's sophisticated computer system is able to guess that a female shopper is pregnant—and in the second trimester, no less.[2]

So how does Target know this?

The company tracks customer purchases and applies sophisticated analytics to the data. It knows based on a purchase of an assortment of 25 items that pregnant women buy. Some of those items include vitamins, zinc, and magnesium supplements. A few extra items are added to the mix—for example, cocoa butter, large clothing, or a rocking chair—and the computer model says the customer is pregnant. Target assigns every customer an ID number that is tied to his or her name, e-mail address, and credit card number.[3] The company is able to track purchases and send coupons based on the customer's preferences. The company even knows that if it can lure mothers-to-be in their second trimester, they will maintain the highest loyalty rates. That is a lot of customer knowledge.

In the case of Target, however, an irate father scolded a Target manager that his teenage daughter received baby coupons in the mail. As it turned out, his daughter was indeed pregnant. The computer was right.

4.1 Driving the Supply Chain

The Sell lever of the supply chain is marketing. We have all heard of recent analytics applications being used by companies to gain insights into customer behavior—from customer loyalty programs to analyzing clickstream data online. However, what does this actually mean for supply chain intelligence? The Sell lever links the organization to its customers (see Figure 4.1).

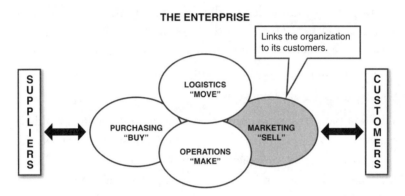

Figure 4.1 The marketing function

Marketing is the function responsible for linking the organization to its customers and is concerned with the downstream part of the supply chain. The task for marketing is to identify what customers need and want, create demand for a company's current and new products, and identify market opportunities. Marketing plays a critical role as providing value to customers drives all actions of the organization and the supply chain.

For an organization and its supply chain to be competitive, it must be better than competitors at meeting customer needs. Marketing is responsible for identifying what these needs are, determining how to create value for customers, and building strong customer relationships. Marketing signals the entire supply chain. To support this, big data analytical applications have been created to capture customer demand, create microsegmentation, and predict consumer behavior. Marketing wants answers to questions such as the following:

- What do customers really want—not just what they say?
- How do we create value in each market segment?
- How much are customers really willing to pay?
- What product features result in Sell—and which ones don't matter?
- How much money should we put into advertising—and how much lift are we getting in return?

Big data analytics on the Sell side answers these questions. However, managing the Sell function is far more complex than it appears. The reason is that customers are driven by much more than their basic needs for products and services. They also often don't understand what drives their own behaviors. Steve Jobs was famous for saying that customers don't know what they want. He referred to a story about Henry Ford and the Model T. Had Henry Ford asked customers what they wanted, the response would have been "A faster horse!"[4]

The key is to understand and develop the combination of products and services that precisely meet the expectations for a cleverly designed set of customers. However, resources are finite and we don't want to spend them on efforts that don't matter. For example, if most customers of athletic shoes only care for three different color choices, offering the product in five colors makes little sense. There would be little benefit and the supply chain costs—extra sourcing, manufacturing, distribution, inventory—would be significant. It is critical to understand exactly what satisfies customers and precisely match product and service offerings.

Marketing is further complicated by the fact that there is no single market for any given product or service. All markets can be broken down into multiple segments, each of which has somewhat different requirements. A critical challenge is market segmentation—in a meaningful and effective way. Segments that are too broad do not permit the company to target a narrow enough group. On the other hand, segments that are too narrow may not be cost effective. *Target marketing* recognizes the diversity of customers and does not try to please all of them with the same product offering.

The challenge for marketing is to identify the most effective market segments and their needs, to coordinate with the rest of the supply chain to make sure production and delivery are feasible, and to ensure that this can be done in a cost-effective manner. After all, marketing can identify an endless array of customer needs. Meeting them, however, may not be financially feasible. It is for this reason that the link to the supply chain is critical. Big data now enables these types of decisions to be made in a holistic manner through comprehensive analyses.

A case in point is a small Napa Valley winery. A few years ago, the marketing group at the winery decided that a smoke-colored bottle would give a distinctive advantage. After some effort, the purchasing group determined that the costs of sourcing would be prohibitive as sources of supply were limited. Most interesting, after some analysis, it turned out that customers were actually indifferent to the bottle choices and that marketing's "hunch" was wrong. These kinds of decisions are now data driving. Had the winery gone with the bottle choice, large financial outlays would have been made for something that really didn't matter.

4.1.1 The Marketing Mix

Marketing provides intelligence that drives the entire organization and supply chain. The intelligence is used to guide the actions of all other functions within the firm, as they modify operations to accommodate new customer preferences. For marketing to be successful, it must be supported by the rest of the organization, as the products customers want must be made available. Remember that customer demand is worthless if there are no products to sell.

Just as the Sell lever drives the decisions of the organization to produce and deliver exactly what customers want, it also drives supply chain decisions. Marketing decisions generally fall into four distinct categories known as the marketing mix or the 4Ps. These decisions are product, price, place (distribution), and promotion. Together, these help marketing identify the right product, sold at the right price, in the right place, using the most suitable promotion. These decisions directly drive supply chain management. Unlike in the past when these decisions have been made speculatively, today most have all been optimized using big data analytics.

4.1.2 Product

Product involves decisions that encompass the bundle of product characteristics that satisfy customer needs. They relate to brand name and functionality. They also address decisions of quality and packaging, which are directly supported by supplier standards and logistics packaging decisions.

Big data analytics has moved product identification to another level. Cross-selling enables correlating customer preferences and understanding the bundles customers buy. Online retailers such as Amazon, Barnes & Noble, and Overstock.com have been identifying such customer product preferences for years. Other companies are taking this capability to the next level by not only identifying correlated products, but also helping customers select products for specific occasions or for select individuals. Sony is using a new technology labeled "online discovery."[5] The system identifies products that may surprise or delight customers. It bases its recommendations on past user behavior and the search terms users have employed. Further, the Web site is integrated to take consumers immediately to the local dealer's Web site where they can experience a highly trained and knowledgeable sales force.[6]

4.1.3 Price

Price refers to the pricing strategy developed for the product. This includes volume and sales prices, seasonal pricing, as well as price flexibility. Although the domain of marketing, pricing strategies and price flexibility are directly tied to supply chain costs. Also remember that this is the only item of the marketing mix that generates revenue. All the others are costs, so you better get this right.

Big data analytics has enabled price optimization for specifically targeted markets and customers. Canadian apparel retailer Northern Group Retail reports that price optimization software has helped it increase gross margins while reducing inventories.[7] Similarly, department store retailer Kohl's credits price optimization with an increase in gross margin.

The effectiveness of different pricing strategies using different marketing mediums for different target groups can now be tested. For example, Kroger uses analytically targeted coupons on which it gets a huge redemption rate, well above the industry average.[8] In addition, Kroger believes the targeted price promotions have increased overall sales. Similarly, the online retailer Overstock.com used an analytics-based gift recommendation system available on its Web site. It found customers who used it bought more than double of those who didn't.

4.1.4 Place

Place deals with having the product where it is needed and when it is needed. These decisions include selection of distribution channel and market coverage. They also include traditional logistics, sourcing, and operations decisions, such as inventory management, order processing, transportation, and reverse logistics. The place component of the marketing mix has become especially complicated as multi-channel order fulfillment becomes a requirement for almost every retailer. Direct-to-consumer is currently the highest growth and highest potential distribution channel. Tracking and managing inventories across multiple channels—and ensuring fulfillment—is one of the biggest challenges.

Adding to the complication is a large variability in demand across channels. Just consider that in the United States shipping volumes for brick-and-mortar stores peak in the fall, while direct-to-consumer demand typically peaks from Thanksgiving through the days before Christmas. Not only are there different channels' needs, but those needs are shifting and evolving. Ensuring fulfillment while keeping costs down is a huge challenge. Estimates are that roughly one fifth of the cost of a product goes into getting the product to the customer. Big data analytics assists in many ways, including analyzing when delivery constraints can be relaxed without negative impacts on customer satisfaction.

4.1.5 Promotion

Promotion is essentially communicating with customers. It deals with advertising and sales techniques to increase product visibility and desirability. Decisions involve promotional strategies and advertising, sales promotions, and public relations. The cost associated with promotion or advertising goods and services often represents a sizable proportion of the overall cost of producing an item. However, successful promotion increases sales so that advertising and other costs are spread over a larger output.

Big data analytics has taken promotion to another level—enabling targeted promotions. Just consider the online discount retailer Overstock.com.[9] The company uses targeted and event-driven e-mail marketing to customize product and pricing offers under different circumstances. Targeting e-mail offers has increased both revenue lift and average order size.

4.1.6 There Is No Average Customer

In the past, companies primarily focused on meeting the demands of the average customer. These companies basically supplied products to customers without trying to determine or satisfy the unique needs of each individual customer. Today, consumers are knowledgeable; they demand what they want at the highest quality, lowest price, and delivered at record speed. Also, today's customers expect a world-class organization to provide high levels of customer service. To compete in this new economy, companies are putting forth a great deal of effort and money to precisely understand their customers, to provide unprecedented levels of customer service, and to provide one-on-one customization. This has resulted in large changes in supply chain management of companies and the relationships between supply chain members who aim to understand and deliver what consumers want.

4.1.7 How Big Data Impacts Sell

Marketing is the lever that has experienced the largest amount of publicity in its big data applications. This is especially true as many applications are designed for tracking consumer behavior and, while providing interesting insights, these applications can also be controversial. Companies using big data analytics along this lever are applying it for customer and market segmentation to drive their supply chains, location-based marketing, sentiment analysis, and in-store behavior analysis. Some are also using it for merchandising, particularly price and assortment optimization. Big data analytics applications have provided more accurate signaling for improved supply chain performance. In fact, increasingly, companies are relying on point-of-sale (POS) data as the demand signal. This is then used to inform raw materials and parts suppliers and drive the entire supply chain. Marketing functions can perform unique kinds of analyses, given proper data applications (see Figure 4.2).

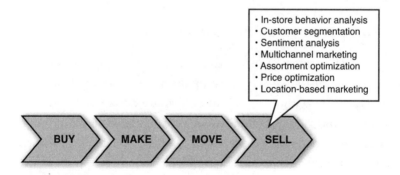

Figure 4.2 Big data applications in marketing

4.2 All About the Customer

As evidenced by Figure 4.2, big data has numerous applications on the Sell lever of the supply chain. Here we expand upon some of these new marketing methods and tools—specifically microsegmentation, cross-selling, in-store behavior analysis, and sentiment analysis.

4.2.1 Microsegmentation

Market segmentation is the essence of Sell, and marketers have been segmenting customers for years.

What is different?

The amount of data available for segmentation has exploded, and the increasing sophistication in analytical tools has enabled the division of customers into ever more granular microsegments. New channels of reaching customers and gathering data—such as mobile devices—are constantly emerging. Data on historical purchases can be combined with behavioral data of individual customers—including clickstream data from the Web—enabling companies to track, segment, and personalize offerings. This is true customer-driven marketing driving the entire supply chain.

Certainly customers are going to be more likely to respond to offers and promotions that are specific to their needs. Companies know this and have been striving to personalize products and target promotions for years. In the past, offers were customized primarily on the basis of demographic segments or clusters—age, gender, income, geographic location—mostly based on syndicated data. More recently, however, companies are focusing on actual customer behavior as a predictor of future buying behaviors. One example is in-store buying behavioral data to target offers. Regardless of the source and type of data, targeting is much more successful with identification of unique customers at the individual or household level. Just recall the case of Target. Therefore, many retailers have established loyalty programs to identify customers or use proprietary credit cards to do so.

Loyalty programs and credit card data have enabled companies to identify and track customers, creating segments and personalizing offerings. These offerings may be based on simple extrapolations of existing customer behavioral patterns—buying and return patterns—or more complex propensity models. Offers may include products with a high likelihood of customer interest. This can easily be computed with correlations between products that are typically purchased in a bundle. Offers can also include targeted promotions, customized product display in catalogs, or on Web pages.

The way offers are delivered has also changed. In the past, offers were often delivered through direct mail. Consumer behavior analysis has indicated these to be less effective. As a result, companies have been delivering targeted promotions through e-mail, Web pages, and in-store kiosks. Tesco, a UK-based retailer, claims to track almost all sales through its ClubCard program.[10] Drugstore chain CVS tracks customer purchases through its ExtraCare card. Data on customer purchase patterns is then used to customize offers provided on register receipts, in kiosks, and in e-mails. Similarly, Hallmark uses its Gold Crown card to segment customers, to customize direct mail offers, and to assess customer reactions to promotions. These loyalty programs have proved invaluable in gathering customer data and creating personalized marketing.

Success of loyalty programs, however, is in part due to the ability of retailers to offer substantial premiums to loyalty customers. This is part of the success of loyalty programs of high-end retailers such as Nieman Marcus and Nordstrom.[11] These companies have sufficiently large revenues per customer to be able to offer such premiums to their most loyal customers. Nieman Marcus, for example, claims that the top customers in its 20-tiered loyalty program account for almost half of its revenues.[12] Top customers can win free fur coats and even a Lexus luxury car. Lower-margin retailers cannot offer these types of premiums. However, they can still take advantage of the benefits of loyalty programs. They can use strategies such as offering targeted coupons and increasingly gathering customer data. Differentiation of loyalty programs will increasingly become important as customers reach a saturation point in their memberships.

4.2.1a Problems

Two major problems exist for companies here. First is the saturation of loyalty programs. Studies suggest that the high proliferation of loyalty programs has resulted in saturation. Most consumers typically carry multiple loyalty cards. In fact, the average household already belongs to more than a dozen loyalty programs.[13] Related to this is the inability for many companies to offer a substantial premium to loyal customers. With so many loyalty programs, consumers are selecting those with the highest rewards. Retailers with high margins are much more able to offer such premiums compared with low-margin retailers.

The second challenge is actually getting customer data. Firms relying on credit card-based identification can typically uniquely identify no more than half of customers. Further complicating matters is the Gramm-Leach-Bliley Act in the United States, which places additional restrictions on uniquely identifying customers based on credit card data.

4.2.1b Solution

A good way to address both of these problems is through collaboration among retailers and supply chain partners. Collaboration can enable combining loyalty programs and can provide sharing of data. For example, Air Miles is the name of a separately operated loyalty program in Canada, the Netherlands, and the Middle East. Points are earned on purchases at participating merchants and can be redeemed against flights with specific airlines. Air Miles tracks spending across a variety of retail and service sponsors; members can redeem the accumulated miles for free flights and other rewards.

Collaboration helps in the sharing of costs. CVS, for example, gathers data on its customers through the ExtraCare program, but the manufacturers of products pay for the coupons. Collaboration can also help in the capturing and sharing of multichannel data, particularly for new and less-traditional channels where data may be difficult to capture.

4.2.2 Cross-Selling

Cross-selling is using analytical approaches to recommend other product offerings for particular customers. We are all familiar with the message "customers who bought what you bought also bought this" when shopping on the Web. The recommendations are typically based on correlations made between what other customers who bought the same product are also buying. This is known as *collaborative filtering*. Recommendations may also be based on specific product attributes. These may actually provide more accurate recommendations. However, a database of product attributes is more difficult and costly to develop if it doesn't already exist within an industry. Delivery of product recommendations is usually online, but could also be delivered by salespeople. The idea is to increase the purchase bundle and it is a very effective tool.

State-of-the-art cross-selling uses all the data that can be known about a customer, including the customer's demographics, purchase history, preferences, real-time locations, and other facts to increase the average purchase size. For example, Amazon employs collaborative filtering to generate "you might also want" prompts for each product bought or visited. The company was a pioneer in this area and began to offer collaborative filtering more than a decade ago. It uses a precalculated item similarity matrix to make real-time recommendations for customers it recognizes through cookies. Amazon has reported that one third of sales are due to its recommendation engine.[14] Another example of this lever is using big data analyses to optimize in-store promotions that link complementary items and bundled products.

Online distribution has made it possible for retailers to offer a much broader range of products. However, it can be difficult for consumers to navigate through a complex product array. Many online retailers have discovered that they can ease this process for consumers by offering recommendation systems. Most retailers have not developed this capability themselves, but have licensed it from software providers. Online entertainment—such as movies and books—is the most popular product category for recommendation systems, but some hard goods retailers use it as well.

These recommendation systems have become very sophisticated. They can recommend purchases for specific occasions or events based on criteria the customer specifies. Overstock.com uses such an application called GiftFinder to help customers find the best gifts for certain occasions. The application asks consumers to specify the occasion, the age of the gift recipient, the relationship to the recipient, and the interests of the recipient. It then recommends specific products. Similarly, Tesco uses a Bayesian system in its online business called Tesco Direct. The recommendation system recommends particular grocery items that customers may want and replaces a previous set of manual recommendations. In addition to Tesco Direct, the company has Tesco Phone Shop and Tesco Mobile, using different channels to reach the customer.

4.2.2a Problem

Cross-selling is increasingly becoming commoditized in online retail. Basically, everyone is doing it. Netflix has built its own recommendation engine, called Cinematch, and has sponsored the Netflix Prize for anyone who can improve its recommendation algorithm. In 2009, the company awarded a $1 million grand prize to the team that created an algorithm that improved Cinematch by 10.6 percent.[15]

4.2.2b Solution

Executives should try to determine how their recommendations will be differentiated from other providers. Differentiation may be difficult, for example, when technology is licensed from providers who sell to multiple customers in an industry.

Recommendation technology is important in online environments, but will be even more important in mobile technology settings, where small screens and lower bandwidth make it more difficult to search through a wide range of products.

4.2.3 In-Store Behavior Analysis

Another way of gathering invaluable information on customer behavior is looking at their in-store behavior. Analyzing data on in-store behavior can help improve store layout, product mix, and shelf positioning. Recent innovations have enabled retailers to track customers' shopping patterns—such as footpath and time spent in different parts of a store. They can draw real-time location data from smartphone applications, shopping cart transponders, or passively monitoring the location of mobile phones within a retail environment.

A growing number of retailers are using sophisticated—and discreet—technology to monitor in-store customer behavior. Many include facial recognition cameras hidden in the eyes of mannequins in order to track shoppers, learn their behaviors, and increase sales.

This type of surveillance system is being utilized by Walmart, called Shopperception.[16] The system links motion sensors to cameras that then track a shopper's product choice on a particular shelf, as well as the time that it takes for the shopper to make a decision. Some retailers use sophisticated image-analysis software connected to their video-surveillance cameras to track in-store traffic patterns and consumer behavior. In-store behavior analysis provides data that can help improve store layout, product mix, and shelf positioning.

Although this can gather valuable information, you should note that it has actually been controversial and may infringe on some privacy issues. In fact, some retailers may leverage *not* using such technologies as part of their marketing campaigns while they gather customer information through other equally effective means. As we have seen in this section, there are numerous opportunities to learn about customers. Investing in all of those opportunities is not possible and avoiding those that are highly controversial may be wise strategically.

4.2.4 Sentiment Analysis

Sentiment analysis is a tool for understanding customer opinions that are expressed on Web sites and in communication centers—from e-mails, forms, surveys, internal files, and reports. It leverages these voluminous streams of data generated by consumers in the various forms of social media to help inform a variety of business decisions. The tool combines statistical learning with advanced linguistics methods to locate and analyze digital content in real time in order to hone in on the sentiment that has been expressed. For example, retailers can use sentiment analysis to gauge the real-time response to marketing campaigns and adjust course accordingly. Just consider that the Gap changed its logo using sentiment analysis and in response to social media in real time.[17] The Gap had put out a new logo using white-and-black coloring and a blue box—an evolution from its old white and blue logo. Using sentiment analysis and watching highly negative social media responses on Facebook, Twitter, and social media blogs, the company "went back to blue" within 48 hours!

4.3 Price Optimization

Essentially, price optimization is using analytics to determine the optimal pricing of products and services along all channels and markets. Historically, pricing has been an art rather than a science. Managers have set prices based on intuition, judgment, and experience. The same was true for setting promotions and markdowns, deciding on which items to offer discounts and how much. The success of these old approaches pales in comparison with the results that are obtained through analytics. Research by Accenture proves this point by finding that roughly one third of retailers using the old approach have at least 10 percent of their merchandise left over at the end of a season.[18] For some, the figure approaches 25 percent. By contrast, companies that implement price optimization typically see huge improvements. New York area grocer D'Agostino Supermarkets found implementing price optimization to yield a rise in gross profit by 16.1 percent and net profit by 2 percent.[19] Similarly, drugstore retailer Duane Reade increased unit sales of baby formula by 14 percent and of disposable diapers by 10 percent in a 20-store trial. The trial was so successful that the company expanded the use of price optimization in all its stores for nonpharmaceutical products.[20]

The analytics part of this is not new. In fact, revenue optimization models have been around for well over a decade. Just consider revenue optimization used for airline seats and then hotel rooms, used by online companies such as Priceline.com. Today, a variety of data sources can be used to optimize pricing decisions in near real time. The increased granularity of customer data on sales and pricing, coupled with the higher computing power, can provide much richer insights. Price optimization software can gather point-of-sale information and seasonal sales data at the store level. This data can then be used to determine probability distributions and create a set of demand curves for each particular SKU in a particular store or cluster location. The demand curves identify the products that are the most and least price sensitive. Additional optimization routines leverage these demand curves to determine optimal recommended pricing. Retailers can use the resulting data to analyze promotion events and evaluate sources of sales lift and any underlying costs that these might entail.

Evidence repeatedly demonstrates that pricing optimization results in an increase in sales and margins. By most estimates, it is how analytics has the greatest impact on the bottom line. In fact, a study by the Yankee Group suggests that returns on price optimization investments typically approach 20 percent.[21] The challenge, however, is that price optimization does normally require a substantial up-front investment. This has been the primary obstacle for small and medium-sized companies. Further, the benefits of price optimization are greatest when undertaken in an integrated fashion with other analytics efforts, such as merchandising optimization, for example shelf space and assortment management. This is clearly the direction in which companies are moving as they look to optimize the entire marketing mix holistically to achieve maximum benefits.

4.4 Merchandising

Merchandising is another element of marketing that has been profoundly impacted—and changed—by big data. Merchandising essentially refers to the variety of products available for sale and the display of those products in such a way that it stimulates interest and entices customers to make a purchase. Two aspects of merchandising are especially impacted by big data analytics. The first is *assortment optimization*. This involves deciding which products to carry in which stores based on local demographics, buyer perception, and other factors. These decisions not only impact sales but have a huge impact on supply chain management. Supply chain management ensures the deliveries of the right quantities at the right location to meet product and quantity requirements—otherwise, there would be no products to sell.

The second aspect of merchandising is *shelf space allocation*. By gathering customer data on the retail store level and observing customer behavior patterns—such as what visually draws or repels the consumer—companies are increasingly learning the types of merchandising strategies that maximize sales.

Merchandising is about making the items available and appealing. How can big data analytics help with that?

Product inventory, shelf space, and displays have always been a valuable resource for retailers. They have competed for the best shelf space such as end displays. Now analytics can be used to determine exactly which merchandising strategies provide the highest level of sales and profits. Think about all the customer behavioral data companies are currently gathering, including in-store behavior analysis. Certainly all this information provides clues as to how best to visually display the items on store shelves.

Historically, assortment and shelf space optimization have been a seasonal activity. With increased competition and awareness of shelf space utilization, it has evolved into a continuous process. Space optimization was often done by manufacturers as category captains in grocery retailing. Today it is evolving toward a joint process between retailers and key suppliers. This is especially true through initiatives such as vendor managed inventory (VMI). Planogramming has increasingly become more computerized with the use of computer-aided design (CAD) tools. However, it has only been through the use of optimization techniques that the real impact on profit of different planograms and store layouts can be realized. Automated planogram generation uses optimization techniques to make store-level tradeoffs between shelf space and assortment depth, while considering desired product display blocking.

Today's analytical tools can analyze the financial impact of different shelf space assortments, the specific preferences of shoppers in the store area, and the likely responses to pricing and promotions. These new tools can provide information on profit contributions of each brand and SKU and even identify cannibalization effects. Early adopters of this technology have been optimizing assortments at the micro level, looking at brands, packaging, and sizing. More companies are now optimizing macro store layouts as well, such as space allocation, and then integrating them with the micro store analysis.

4.5 Location-Based Marketing

Location-based marketing is a new and rapidly evolving big data application along the Sell lever. It is based on personal location data—such as mobile devices—and can enable real-time responsiveness to customers. It can then be used to target consumers who are close to stores or already in them. As a consumer approaches a store, for example, that store may send a special offer for an item to the customer's smartphone. In fact, it is estimated that nearly 50 percent of smartphone owners use or plan to use their phones for mobile shopping.[22] The company PlaceCast—specializing in location-based marketing—claims that more than 50 percent of its users have made a purchase as a result of such location-based ads called ShopAlerts.[23]

There is an explosion in the amount of information available about where people are in the world. An early source of personal location data came from individuals' credit and debit card payments. This information was linked to personal identification data from cards swiped at point-of-sale (POS) terminals, which are typically in fixed locations. The advancement of technologies such as GPS allowed us to quickly locate a device as small as a mobile phone within a few dozen meters. This type of personal location data can pinpoint accurately within a few city blocks where a person—or really the device—is in real time.

4.5.1 Geo-Targeted Advertising

Geo-targeted mobile advertising is one of the most common ways organizations can use personal location data. Consumers can sign up to receive geo-targeted ads. Personalized ads can then appear on the consumer's smartphone when he or she is close to that store. This could also include coupon offers from restaurants or coffee shops. These geo-targeted advertisements appear to be far more effective than traditional forms, such as TV or print. The reason is that geo-targeting offers the ads in real time precisely when the purchase decision is about to be made. This is a fast-growing area of analytics application and companies are investing increasing rates in this service for advertising.

4.5.2 Tracking Shoppers

Personal location data has tremendous value in understanding customer shopping patterns. Technologies such as GPS and RFID enable tracking shoppers to identify buying patterns. None are perfect and each has certain benefits. GPS signals, for example, often do not penetrate indoors so companies have to use other technologies to track shoppers. One option is RFID with tags placed on shopping carts. Today, RFID tags are cheap and accurate. They enable companies to monitor customer movements in real time as they move about the store. The problem is that they monitor the cart and not the shopper, so often they provide wrong information. Shoppers, for example, often leave carts in aisles while they go about their shopping.

Other options are video cameras. They can be an excellent source of information for understanding general traffic flow. However, they can be difficult to use for following an individual shopper's behavior. Newer technologies are being refined, as this is such a fruitful area for understanding shopper behavior.

Simply looking at customer location information can tell companies about foot-traffic density—where the majority of customers are going. It also provides detailed insights about where shoppers slow down—such as looking at a particular type of display. This data can be studied to see responses to promotions and advertising, and then these patterns can be linked with data on product purchases, customer demographics, and historical buying patterns. This type of detailed information provides intelligence to offer better product choices, better product assortments and merchandising, and improved store layouts.

4.5.3 The Customer Is Where?

As the availability of personal location data becomes more common and awareness of its value more widespread, new business opportunities are emerging that take advantage of both aggregated and raw data. One example is an application called Street Bump launched by the city of Boston.[24] The app takes advantage of personal location data to detect potholes. Street Bump uses technology already built in to smartphones, including GPS and accelerometers. It then notes the location of the car and the size of the potholes it crosses. To make things even more interesting—and improve on an already good thing—the city has issued a public challenge to users to find the best methodology for mapping street conditions and making Street Bump as useful as possible.

Another example of a new business value is through the analysis of aggregated location data. Sense Networks is a company that uses real-time and historical personal location data for predictive analytics. Its first application for consumers was CitySense, a tool designed to answer the question: "Where is everyone going right now?"[25] City-Sense shows the overall activity level of the city. It shows hot spots and places with high activity, all in real time. The tool then links to Yelp and Google to show what venues are in those specific locations. Another application from Sense Network is called CabSense. It offers users an aggregated map that ranks street corners by the number of taxicabs picking up passengers every hour or every day of the week. That is information that everyone can use.

4.6 The Whole Bundle

In this concluding section, we put together the big picture from this chapter. The bottom line is: big data analytics can tell an organization how to spend its marketing budget, and allow a company to utilize various marketing methods and tools simultaneously to provide a holistic picture of the consumer.

4.6.1 Where to Spend?

An old complaint goes like this: "Half of the money I spend on advertising is wasted; the trouble is I don't know which half."[26]

With big data analytics, this is no longer true. To achieve the greatest impact on the marketing mix, companies have many possible media and promotional approaches to choose from. The number of possibilities is also growing all the time as technology rapidly advances. Further, companies can optimize the marketing mix—product, price, place, and promotion—together in one bundle rather than as separate entities, determining which marketing investments work and which are less effective.

Resources are finite. Marketing mix models help companies determine where they should spend their marketing resources for maximum impact. Marketing mix models allow companies to determine which marketing vehicles provide the greatest sales lift. Using econometric models, they can determine how much lift a particular promotional approach generates for which target market. Variations and combinations of variables can be tested for maximum impact. It may be, for example, examining how often a particular campaign should be run, with what frequency, and through which medium.

The rise of marketing mix models is one factor behind the decline of traditional print and broadcast advertising, and the rise of online media. In general, marketing mix analyses have determined that the traditional, nonaddressable media are less effective overall at lifting sales, and retailers and other advertisers have responded by reallocating their resources. For example, department store chain Macy's undertook a detailed analysis of its marketing mix. Results were clear. Spend less on print and more on other media. More resources were freed up for advertising, more marketing was devoted to local markets, and guidelines have been provided to local marketers on the ideal marketing mix for their markets. Similar outcomes were uncovered by office products retailer Office Depot when they embarked upon multiple analytical projects. An analytical analysis of the media mix used 36 models that analyzed channel, segment, and department sales data from three years of transactions.[27] The lesson was to use circulars sparingly and the space in the circulars had to be earned by product categories through demonstrated sales.

4.6.2 Together Not Separate

Numerous applications exist for companies to optimize elements of the marketing mix—from understanding customer behavior to price optimization to merchandising. However, these elements are all related. Merchandising affects pricing, which affects store behavior. What is really new and important is to optimize these things holistically as a bundle and not separately.

An excellent example is that assortment and space optimization are also closely linked with price optimization. It is difficult to optimize prices on the wrong product inventory. Further, assortment and shelf space optimization impacts one of the broadest sets of business processes among the categories available to retailers. Apparel retailer Limited Brands views assortment optimization as highly significant to sales.[28] It uses a randomized testing approach and control groups to improve store assortments, as well as segmentation of stores by brand and customer.

Because these decisions are highly linked, executives should develop a plan and sequence for how they will implement such a broad capability. It is also very important to coordinate implementation of these capabilities with suppliers and other supply chain partners. Further, assortment plans must be developed considering replenishment and packaging—all supply chain decisions.

An added complication is the growth of multiple channels and online demand. The growth of the *omni-channel* has been especially challenging for retailers. This involves the seamless interface of the different channels to create one unifying experience for the customer. For the customer, this means availability, ease, and the ability to purchase across channels. For the retailer, this involves huge coordination of inventories, variations in demand, and pricing. Companies are aware that in order to drive sales, customer satisfaction, and loyalty, they need to enhance the omni-channel experience for consumers. Retailers can use big data to integrate promotions and pricing for shoppers seamlessly, whether those consumers are online, in-store, or perusing a catalog. Williams-Sonoma, for example, has integrated customer databases with information on some 60 million households.[29] The company tracks information such as household income, housing values, and number of children. It then uses this information to target e-mails. The response based on this information is up to 18 times the response rate of e-mails that are not targeted.

It is important to remember the linkages between the elements of the marketing mix. These linkages further need to be tied to the other supply chain levers. The advances in big data analytics provide an unprecedented view of consumer behavior. Marketing, however, needs to work with operations to ensure production of the product with the exact characteristics needed; sourcing needs to ensure that supplies are available in a timely and cost-effective manner; logistics needs to ensure deliveries of exactly what is needed throughout the supply chain; also, finance needs to be involved in securing appropriate funding. These advances in marketing don't mean anything if they are not linked to fulfillment. This involves coordination across the supply chain levers.

5

Impact on "Make"

Manufacturing or data problem?

Just a few years ago, Dell customers could order customized computer configurations on the company's Web site using the configure-to-order model.[1] This included variations in models, software configurations, memory, screens, design, and every other customizable feature. Although great for customers, this resulted in more than seven septillion[2] possible configurations of Dell products. Think about the inventory implications.

Then, the company moved to an analytics-driven system. They wanted to optimize inventory decisions to satisfy the diverse needs of a broad set of customers but also keep costs as low as possible. The analytics team decided to use historical order data to run cluster analysis to determine the most common configurations customers were choosing. It turns out there was a lot of commonality in ordering. Going through the data, they were able to strip down the seven septillion options to just a few million. They were then able to identify certain models so common that the company could stock them in pre-configured inventory just waiting to be shipped. These common configurations could then be built ahead of time with the lowest margins. Ordered today, the customer can then have them tomorrow. That is high customer service with low inventory cost. The new system has brought the company an extra $40 million in positive revenue. This allowed the company to Sell exactly what the customer wants but Make it in a cost-efficient manner.

5.1 Making the Things We Sell

Make is the supply chain lever that is operations management. Operations management is the business function responsible for producing a company's goods and services. It is responsible for orchestrating the efforts needed to transform inputs into finished products. Without operations, there would be no products to sell—whether tangible goods or services (see Figure 5.1).

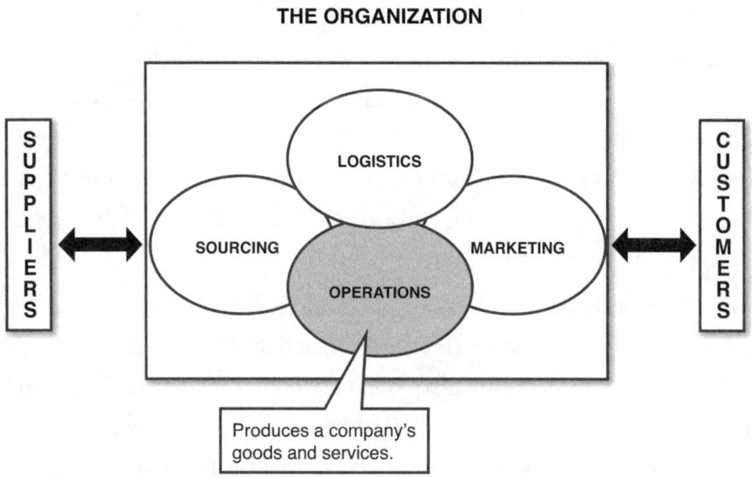

THE ORGANIZATION

Figure 5.1 The Operations function

Operations management is responsible for planning, organizing, and managing all the resources needed to produce a company's products. This includes people, equipment, technology, materials, and information. It is a core function of every company, whether large or small, service or manufacturing, profit or nonprofit. Consider The Gap Incorporated. Gap's marketing function identifies characteristics of products desired by customers for each of its three brands—Banana Republic, Gap, and Old Navy. However, it is operations management that must design and produce the merchandise, from T-shirts to jeans. It must also ensure that the right products—variety, colors, and styles—are available in the right quantities at various retail locations across the globe. And it must be done in a cost-effective manner. Operations management can be described as playing a transformational role in the supply chain by transforming resources into finished goods and services (see Figure 5.2).

Figure 5.2 The Transformational role of Operations

5.1.1 Optimizing Operations

Operations management is complex. There are many different decisions that are involved in transforming inputs into finished products. For operations to be able to design and produce the products customers want, customer preferences—such as likes and dislikes—have to be translated into tangible characteristics. It is then up to operations to decide on the exact type of process needed to produce the desired product. This includes decisions on equipment, level of technology and automation, skill levels and job requirements of workers, and capacity of facilities.

Understanding the scope of these decisions is important as it helps us know which questions we need to ask with big data analytics. Big data analytics has had a tremendous impact on improving decisions along the Make lever—from impacting product design to quality management (see Table 5.1). It is important, however, to focus on the applications that address the important issues for the organization. Anything can be optimized. The question is: Does it matter?

Table 5.1 Operations Management Decisions

Decision Area	Responsibility
Product design	Designing unique product features
Process design	Creating the production process to produce the designed products
Quality management	Establishing and implementing quality standards
Inventory management	Deciding on amounts of inventory to carry and when to order
Facility layout	Deciding the physical layout of the production facility
Facility location	Deciding on the best location for facilities
Scheduling	Creating schedules for workers, machines, and facilities

Intense global competition has placed even greater importance on the Make lever of the supply chain. Organizations within the supply chain network must operate at the highest level of efficiency and utilize the most current operations methods. In fact, many inroads that are made in competitive markets come from innovations in operations, such as Tata Motors producing less-expensive vehicles due to new production techniques. Big data analytics applications have enabled optimizing many of these decisions, substantially improving efficiency.

5.1.2 Coordinating Make and Sell

Without Make, there would be no products to Sell. However, operations cannot succeed without coordinating with other business functions. In particular, operations must work with the Sell lever to understand and design the exact products customers want. Make must then create the production processes to efficiently produce these products. This is where big data analytics capabilities—such as microsegmentation, sentiment analytics, customer behavior analysis, and location-based marketing—come in. These new capabilities move understanding of customers from hunches to a fact-based understanding. We no longer have to guess what the customer really wants.

The Sell lever must also understand the capabilities of Make, such as the types of products it can—and cannot—produce. This may sound trivial, but it is very important. Otherwise, companies will Sell products the company can't Make. Consider a furniture manufacturer in North Carolina where marketing pushed for the selling of laminated versus customized furniture, only to find that the operations requirements were completely different for one versus the other. The company was simply not able to deliver. Based on a thorough understanding of the Make and Sell levers, an organization's supply chain manager must coordinate Make and Sell activities (see Figure 5.3).

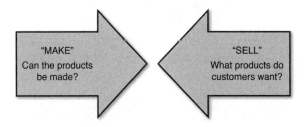

Figure 5.3 Coordinating Make and Sell

It is not enough for a company to manage its own operations function. These must be coordinated across the supply chain. The reason is that each company depends on other members of its supply chain to be able to deliver the right products to its customers in a timely and cost-effective manner. Processes operate as a system and every system has a constraint—a bottleneck—that must be managed.

In the upstream part of its supply chain, a company depends on its suppliers for the delivery of raw materials and components in time to meet production needs. If deliveries of these materials are late, or are of poor quality, production is delayed, regardless of how efficient a company's operations process. On the downstream side, a company depends on its distributors and retailers for the delivery of the product to the final customer. If these are not delivered on time, are damaged in the transportation process, or are poorly displayed at the retail location, sales will suffer. Also, if the operations function of other supply chain members is not managed properly, excess costs will result, which will be passed down to other members of the supply chain in the form of higher prices.

5.1.3 Both Manufacturing and Service

Manufacturing organizations were early users of analytics applications to drive quality and efficiency. Since the dawn of the computer era, manufacturing firms were using information technology and automation to design, build, and distribute products. In the 1990s, manufacturing companies achieved large productivity gains due to new operational improvements. These were focused on increasing operational efficiency through lean systems and improvements in quality such as Total Quality Management (TQM). These advancements in manufacturing were quickly followed by applications in service organizations that began to achieve similar benefits. The same is true today with applications of big data analytics. However, these applications need to be modified to the unique needs of services.

Manufacturing and service organizations each pose unique challenges for the operations function. There are two primary distinctions between them. First, manufacturing organizations produce physical, tangible goods that can be stored in inventory before they are needed. By contrast, service organizations produce intangible products that cannot be produced ahead of time. Second, in manufacturing organizations, most customers have no direct contact with the operation. For example, a customer buying a bottle of Pepsi never sees the bottling factory. However, in service organizations the customer is typically present during the creation of the service, such as at hospitals, colleges, theaters, and barber shops.

The differences between manufacturing and service organizations are not as clear-cut as they might appear, and there is much overlap between them. Most manufacturers provide services as part of their business, and many service firms manufacture physical goods they use during service delivery. For example, a manufacturer of furniture may also provide shipment of goods and assembly of furniture.

A barber shop may sell its own line of hair care products. Even in pure service companies, some segments of the operation may have high customer contact while others have low customer contact. The latter can be thought of as "backroom" or "behind the scenes" segments. Examples would include the kitchen segment at a fast-food restaurant, such as Wendy's, or the laboratory for specimen analysis at a hospital.

It is important to remember that strong big data applications exist for both manufacturing and service organizations, although some are more applicable to one industry versus the other. As tangible goods become more commoditized, manufacturers are increasingly developing integrated product-service offerings that differentiate products and deliver additional value. This is called servitization.[3] Servitization examples in industry abound, such as Rolls-Royce selling not just aircraft engines but complete service solutions that include all operations and maintenance functions.[4] Similarly, tool manufacturer Hilti expanded its offering from the products to the broad service capabilities that the tools deliver.[5] These companies will need a broad set of applications that address both the tangible and intangible aspects of Make.

5.1.4 How Big Data Impacts Make

Companies are using analytics along the Make lever to help improve every aspect of their operations—from quality to efficiency to labor utilization. Western Digital, one of the world's largest computer hard disk drive manufacturers, for example, uses analytics to achieve the highest levels of quality in its industry.[6] To achieve this level of quality, the company has transformed its manufacturing process to track product quality throughout the production process. The system allows scanning, recoding, testing, and tracking of every disk drive it produces while they are still on the line. The company runs analytics in real time on the production floor to help them identify and pull bad disks long before they reach the customer. Even if a disk passes initial review, the testing continues. If further analysis reveals a problem, it can be located and pulled from inventory bins. This has resulted in a rate of 1.9 percent defect per million, which was the lowest warranty return rate in the industry in 2010.

Like Western Digital, a large number of companies report using big data analytics for inventory management, optimization of stock levels, maintenance optimization, and some in-facility location.[7] Some are considering use in the workforce productivity evaluation as well as study of capacity constraints. Productivity and quality have huge applications. With the help of big data analytics, companies can now run daily analyses of operational performance. These statistics can be aggregated and reported by location, SKU, department, and even employee. These systems are now moving ever closer to real time where they can monitor quality and alert companies of defects and other problems. Although companies have been using technology to evaluate performance for many years, what is different is timing and scale. Today's technology enables real-time monitoring at the highest level of detail, and many of these decisions can be automated into a "digital factory."

Big data analytics impacts the Make lever by enabling higher efficiency in product design and production, improvements in product quality, and better ability to meet customer needs through more precisely targeted products. Big data can help manufacturers reduce product development time by 20 to 50 percent and eliminate defects prior to production through simulation and testing, according to estimates.[8] Efficiency gains can be achieved across the entire supply chain, from reducing unnecessary iterations in product development cycles to optimizing the assembly process to maximizing the use of labor. The company is producing the exact products customers want in the most efficient way possible. That is efficiency and effectiveness in tandem.

Labor optimization is another operational lever that is seeing large applications of big data analytics. Companies can now reduce costs and maintain service levels through measures such as optimizing labor, automating and tracking employee attendance, and improving labor scheduling. For a summary of the applications of big data in operations management, see Figure 5.4.

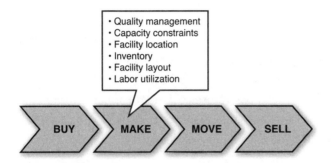

Figure 5.4 Big data applications in operations

5.2 Product Design and Innovation

Big data analytics can impact the Make lever not only along tactical decisions, such as identifying inventory stock levels, but it can also answer the threshold question of what exactly the company should make in the first place. It can help foster innovation by building cross-functional research and development (R&D) and product design databases along the supply chain to enable concurrent engineering, rapid experimentation and simulation, and co-creation of products.

5.2.1 Product Design

Product design is the process of specifying the exact features and characteristics of a company's product. Ideas for which features to include are driven by the Sell lever—marketing—whose job is to understand what customers want. Marketing understands that customers respond to a product's appearance, color, texture, and performance, as well as packaging, delivery, and service. Sell and Make then work together to create the product bundle that meets these consumer preferences.

Product design decisions are one of the most important in operations management as they drive other operations decisions. One of these is *process design*. Process design is the design of the production process needed to produce the desired product. This includes decisions regarding technology and automation, implementation of lean systems, and performance measurement. Product design defines a product's characteristics and then translates them into measurable dimensions the process can use to produce the product. Product and process design affect product quality, production cost, and customer satisfaction. Do you need Six Sigma and does the customer care?

Consider Apple's iPhone 4 manufactured in 2010. The product design specified white to be one of the color options. However, after white iPhones were produced, consumers found the white casing to "leak" too much light from the phone unit, washing out pictures and creating other problems. The white idea was great. The production process, however, could not deliver. If the product is not well designed, or if the production process is not capable of producing the exact product design features, the quality of the product will suffer. Therefore, for a product to succeed in the marketplace, it must have a good design *and* a good production process. These decisions are all intertwined.

Where does big data come in?

5.2.2 *What the Customer Wants*

Historically, product design decisions were made based on guesswork and instinct, using traditional marketing tools to identify what customers wanted. The use of big data analytics changes this. It offers opportunities to improve and substantially accelerate product design in a number of ways, both in manufacturing and service. First, big data on the Sell side helps identify the most important and valuable features based on concrete customer inputs rather than hunches. Analytics can identify designs that meet customer criteria while minimizing production costs. It can also harness consumer insights to reduce development costs through processes such as open innovation. Big data can improve efficiency and effectiveness of the design process, increasing the value-added content of products and services at a lower cost.

Obtaining customer input through market research has traditionally been a part of the product design process—a process that results in mere estimates. The tools available to the Sell lever are helping change this. However, most companies have yet to systematically extract crucial insights from the increasing volume of customer data. What we need are mechanisms to use this data to refine existing designs and help develop specifications for new models and variants.

One tool that can help and is increasingly being used by best-in-class companies is *conjoint analyses*. This is an analytical tool that can determine how much customers are willing to pay for certain features and to understand which features are most important for success in the market. Leading-edge companies then supplement such efforts with additional quantitative customer insights mined from sources such as point-of-sale (POS) data and customer feedback. New sources of data that manufacturers are starting to mine include customer comments in social media and sensor data that describe actual product usage.

Collectively, these efforts provide insights that more precisely drive product design. However, gaining access to comprehensive data about customers to achieve holistic insights can be a significant barrier. The reason is that many companies—from distributors, to retailers, to other players—are often unwilling to share such data. They consider it a competitive asset. Nevertheless, the value achieved through successful design can be substantial.

5.2.3 Service Design

Big data can greatly enhance product design of services. Product design in the service industry has an added complication as the product is intangible and there is a high degree of customer interaction. Service design requires the design of the entire service concept. As with a tangible product, the service concept is based on satisfying customer needs. However, there are aesthetic and psychological benefits the service provides that must be considered. Services are "experienced" by the customer, and details that enhance this experience are all part of the service concept. An example of this is seen in the retail sector with the trend of Shopper Marketing.[9] Here, retailers focus on creating the entire shopping experience, rather than just focusing on features of one product. Capturing and analyzing data on customer behavior is essential for this.

Companies can leverage data to design products that better match customer needs. Data can even be leveraged to improve products as they are used. Consider an example of smart mobile phones we are now seeing. The phone "learns" its owner's habits and preferences. It holds applications and data tailored to that particular user's needs. As a result, it is more valuable than a new device that is not customized to a user's needs.[10] Also, it is now becoming economically feasible to embed sensors in products that can "phone home." These devices can then serve to generate data about how the products are actually used. Manufacturers can now obtain real-time input on how products are used and monitor for emerging defects. They can then respond immediately by making changes to the production process. R&D and operations can share this data as they engage in redesign and new product development.

5.2.4 Co-Creation

Companies have used computerized technologies to manage aspects of product design and manufacturing for many years. Examples include computer-aided design, engineering, manufacturing, product development management tools, and digital manufacturing. Unfortunately, the large data sets generated by these systems have tended to remain isolated. Integrating these data sets that reside in independent systems can enable effective and consistent collaboration and co-creation. This type of co-creation would reduce development time, improve quality, and reduce resources. Dismantling these "silos" of information is a challenge as companies adopt and implement new technologies.

Co-creation is particularly useful in industries where a new product is typically assembled with hundreds of thousands of components supplied by hundreds of suppliers from around the world. Here, having an original equipment manufacturer (OEM) co-create designs with suppliers can be extraordinarily valuable. It can enhance design, testing, and experimentation. Designers and manufacturing engineers can share data. They can quickly create simulations to test different designs, test the choice of parts and suppliers, and compute the associated manufacturing costs. This is especially useful because decisions made in the design stage typically drive 80 percent of manufacturing costs.

This approach has been shown to be effective in both the aerospace and auto industries. Toyota, Fiat, and Nissan have all used collaborative design and data experimentation methods. As a result, they have all cut new-model development time by 30 to 50 percent; Toyota claims to have eliminated 80 percent of defects prior to building the first physical prototype.[11]

5.2.5 Innovation

Big data enables innovative services and even new business models in manufacturing. Sensor data has made possible innovative after-sales services. For example, BMW's ConnectedDrive offers drivers directions based on real-time traffic information, automatically calls for help when sensors indicate trouble, alerts drivers of maintenance needs based on the actual condition of the car, and feeds operation data directly to service centers.[12] The ability to track the use of products at a micro level has also made possible monetization models that are based not on the purchase of a product but on services priced by their usage, as described previously.

Companies are increasingly relying on outside inputs to drive innovation and develop products that address emerging customer needs. They are also using innovative channels to access this talent. This is called *open innovation*.[13]

With the advent of Web 2.0, some companies are inviting external stakeholders—consumers, suppliers, and other third parties—to submit ideas for innovations or even collaborate on product development via Web-based platforms. Consumer goods companies such as Kraft and Procter & Gamble (P&G) invite ideas from their consumers. They also collaborate with external experts in new product development, including academics and industry researchers. It is estimated that half of the new products at P&G now have elements that originated outside the company.[14]

Open innovation through big data analytics has been extended to other industries as well. BMW, for example, has created an *idea management system* to help evaluate ideas submitted through its "virtual innovation agency."[15] This has cut the time taken to identify high-potential ideas by 50 percent. It has also reduced the time in assessing the feasibility of an idea. This process has resulted in the incorporation of two or three major ideas from its open innovation effort into its models every year.

5.3 Improving the Production Process

Inventory management, quality management, and labor utilization are three major areas where big data analytics can bring about improvements in the production process.

5.3.1 Inventory Management

Inventory management is one of the most important decision areas for Make. Not having the right goods when and where they are needed means lost sales and poor customer service. Too much inventory means high inventory carrying costs. In fact, carrying costs can be as high as 50 percent of item costs.[16] Managing this tradeoff is a constant problem for Make.

Inventory management has one of the largest applications of big data analytics. RFID tags, for example, are now affixed to items and allow tracking and monitoring of stock levels, location status, in-transit shipments, and can trigger automatic reorder quantities. These sensors enable real-time visibility for in-transit inventory. They also enable monitoring environmental conditions, such as temperature control for perishable items, as well as providing a time stamp on how long inventories spend in different stages of transit. Coupled with POS data on the Sell side, they can create intelligent inventories that know exactly how much stock needs to be placed in a specific location and automatically generate orders.

With big data applications, we sometimes hear the term *adaptive* inventory. This is where companies engage in real-time inventory tracking and are able to identify stocking levels of any item at any location. The inventory levels essentially *adapt*. By combining multiple data sets, such as sales histories, POS data, weather predictions, and seasonal sales cycles, forecasting accuracies are improved. Algorithms then automatically connect forecasts to current stock levels at a granular level. This results in optimized inventories.

Big data applications permit supplies to be more tightly coupled with demand signals—called *demand sensing.* Sensing the real-time demand signal can make inventory more "productive." Here, lower levels of stock can offer higher customer service levels. Real-time location sensing can create a better supply-demand match. This highly accurate information can feed into all levels of inventory calculations, from production planning levels to stocking levels to ordering levels.

5.3.2 Quality Management

Big data analytics substantially changes and improves quality management. Total Quality Management (TQM) is an integrated organizational effort designed to improve quality at every level of the organization. Quality management essentially requires monitoring quality to identify defects when they occur.

Two significant aspects of quality management are changed with big data analytics. First is the newfound ability to precisely identify customer-defined quality. A key tenant of quality management is meeting quality expectations as *defined by the customer.* Historically, companies used traditional marketing tools, such as surveys and focus groups, to try to understand what dimensions of quality the customer considers most important. This analysis was called The Voice of the Customer. This type of guessing, however, is becoming a thing of the past.

All the Sell tools discussed in Chapter 4—from customer microsegmentation to in-store-buying analysis—allow companies to know what customers are actually buying with great accuracy. There is no need to try to guess what customers want. We can measure in real time what customers are buying and use sophisticated analysis to quickly determine what drives their behavior. This information is then used to inform Make, Move, and Buy with great precision.

Second, traditional quality management focuses on identifying quality defects as early as possible in the production process. TQM tools have traditionally included the *seven tools of quality.*[17] These tools include checklists, fishbone diagrams, histograms, Pareto analysis, scatter diagrams, flowcharts, and control charts. Control charts, for example, rely on random sampling to test whether a process is operating within expectations for particular dimensions (e.g., product weight, dimensions, or number of customer service complaints). These tools are meant to be used in tandem to help companies detect and root out causes of quality. These processes, however, are quickly becoming ancient history.

Why?

The reason is that we now have sensors and tags attached to products that can monitor performance in real time. There is no need to do randomized product sampling. Yes, sensors are costlier but also much more accurate. They are also rapidly becoming affordable for everyday use. Consider that readings from a heat sensor on a factory machine can detect problems before they occur. Sensors can detect that an engine is prone to a breakdown based on the heat or vibrations that it produces. Sensors can be affixed to bridges and buildings to watch for signs of wear and tear. Sensors are also used in large chemical plants and refineries, where a piece of broken equipment could bring production to a standstill; the cost of collecting and analyzing the data that indicates when to take early action is lower than the cost of an outage. All this changes the traditional concepts and standards of quality.

Honda is an excellent example of how companies can use big data to impact quality. The company has developed software to provide an early warning system that identifies major quality problems in its vehicles before they become catastrophic.[18] The company uses analytical software to mine three sources of data—warranty service records categorizing quality problems (e.g., brake system, headlamps), free text notes from mechanics, and transcripts of phone calls from customer to call center—to identify trends or emerging problems. These are then forwarded to human analysts to make risk management judgment calls. This type of system also illustrates how big data analytics can enhance and improve human decision making.

How about Volvo? The company is also using analytics to improve quality.[19] By 2020, Volvo is shooting for a perfect quality record. This means zero accidents, zero deaths, and zero injuries in any of its cars. To accomplish this goal, the company has turned to big data analytics. Volvo has arranged a process where every time a car pulls into any of its dealerships for any reason, a technician downloads all of its sensors and all of the information it has been accumulating since its last service. That information is then sent to Volvo headquarters and added to its expansive data warehouse. The amount of data accumulated is huge—containing all of the diagnostic data from all of its cars for the past six years—and adding up to somewhere more than 1.7 terabytes of data to date. With that data, the company can have early detection—and correction—of any kind of problems. This has so far reduced the time to correct defects from as much as eight months to as little as three weeks.

The key here is that the old notions of quality are rapidly changing, from the ability to precisely know customer-defined quality to identifying defects in real time.

An increasingly important application is using sensor data from products once they are in use to improve quality and service offerings. For example, analyzing the data reported by sensors embedded in complex products enables manufacturers of aircraft, elevators, and data-center servers to create proactive, smart, preventive maintenance service packages. A repair technician can be dispatched before the customer even realizes that a component is likely to fail. Other manufacturers have been able to use this capability to transform the commercial offering from one of a tangible product to one that is focused on a service. Good examples are traditional manufacturers of jet engines such as Rolls-Royce, General Electric, and Pratt & Whitney who now sell "power-by-the hour" and such similar services.

5.3.3 Labor Utilization

Labor is often the most significant expense after cost of goods. For many labor-intensive organizations—retailers and distribution centers—workforce scheduling is one of the most pressing problems. Tools for labor scheduling can not only reduce management time spent on this task, but can also do a better job of scheduling workers with the right skills when customers are most likely to visit their stores. The most advanced systems use sophisticated forecasting to consider changes in demand due to upcoming promotions, arriving inventory, and local macroeconomic data. New software has the ability to consider the most productive employees. These workers can be scheduled during peak sales hours in retail or have the highest value orders routed to them in a warehouse. Including these types of factors can result in significant cost savings from workforce efficiency.

Walmart, for example, has implemented a workforce management system that schedules workers based on predictions of when customers will be most likely to shop. The approach is currently being adopted by other retailers as well. One criticism of the software is that it has assigned workers to split shifts and hours that they don't want.[20] Employee preferences are important as they impact productivity and newer applications are taking this into account.

Other companies are using applications that take into account employee preferences. Outdoor retailer Recreational Equipment, Inc. (REI) is using a workforce management application that supports labor forecasting and workforce optimization.[21] The company estimates that the resulting improvements in customer service have led to 1 percent higher sales. Employees generally feel that the system leads to greater equity in scheduling.

Demand forecast accuracy is critical for an efficient labor schedule. This is yet another example of where data captured on the Sell side needs to inform Make decisions. The increasingly sophisticated forecasts on the Sell side need to be synchronized across the organizational plan. Otherwise, companies will not be able to take full advantage of those new labor-scheduling tools.

5.4 The Digital Factory

This is an exciting time for Make. A wave of new technologies is emerging, maturing, and converging in a way that is reshaping product design and manufacturing, shifting from a world defined by hardware and traditional production constraints to one that is largely defined by software and data. The increasing deployment of the *Internet of Things* is allowing manufacturers to use real-time data from sensors to track parts and monitor machinery, software, and analytics to guide and completely reshape operations.

5.4.1 Transformative Technologies

Recent research conducted by IBM has identified three critical and emerging technologies that are transforming manufacturing: three-dimensional (3D) printing, a new generation of intelligent assembly robots, and the rise of open source hardware.[22] Individually, each of these trends is transformative. However, together—and driven by big data analytics—their power is multiplied.

The first of these trends is 3D printing—a technology that promises to change manufacturing as we know it. Three-dimensional printers use technology that is similar to ink-jet and laser-jet printers. They deposit materials like plastics and metals in thick layers one atop the other. The process gradually builds up one layer at a time until the object is produced. This literally means that a solid object can be created from a software design with just a click of a button. This technology makes desktop manufacturing possible and it requires no economies of scale. Just a few years ago, three-dimensional printers were considered useful tools in prototype studios. They are now rapidly becoming essential machine tools on the production line. Just consider what opportunities that offers in manufacturing flexibility. Imagine coupling that with real-time demand data and we are changing the Make lever as we know it.

A second key emerging technology is a new generation of intelligent assembly robots. Past robotic systems required huge, complex installations. These typically started around \$250,000 per assembly station, well above the cost of what most manufacturers could afford. This new generation costs a fraction of that price—at around \$25,000 per robot.[23] They can also be installed in less than a day. Suddenly, efficient, effective manufacturing automation is within the reach of even small companies. Coupled with big data algorithms to drive performance, superproduction capability is possible.

The third emerging transformational technology is the rise of open source hardware. Open source hardware is electronic or computer hardware built from design information that would have traditionally been copyrighted or licensed. Instead, it has now been made available for use at no charge. Such information can include documentation, schematic diagrams, construction details, parts lists, and logic designs. This is all information needed to build the product—no reverse engineering required. Initially, this documentation was accessible to hobbyists and the very smallest businesses. Now this capability has spread into the realm of broad hardware design. From mechanical systems to networking equipment, hundreds of product designs are now available to anyone. This includes items from personal computers to computer peripherals to telecommunications.

Imagine the collective potential of these technologies coupled with big data analytics. Collectively, these technologies are removing all the physical constraints of the Make lever—building molds, ordering parts, reconfiguring assembly machinery—and turning them into processes that can be run through software. This is called the *software-defined supply chain.* Driven by big data analytics, it becomes the digital supply chain.

5.4.2 Sensor-Driven Operations

Taking inputs from product development and historical production data (for example, order data, machine performance), manufacturers can apply advanced computational methods to create a digital model of the entire manufacturing process. Such a *digital factory*— including all machinery, labor, and equipment—can be used to design and simulate the most efficient production systems. For any specific product, the simulation can create the facility layout and sequencing of steps. Leading automobile manufacturers have used this technique to optimize production of new plants. This is particularly valuable as there are numerous constraints that cannot be considered manually, from capacity to labor.

An example of how high-sensor monitoring can improve quality is seen at BP's Cherry Point Oil Refinery in Blaine, Washington.[24] Managers use big data analytics to detect trends in oil corrosiveness, which can be highly damaging to pipes. The company uses multiple sensors across the plant to generate vast amounts of information. This serves to compensate for "data fuzz" created by the intense operating activity in the environment (such as noise, heat, vibrations). This is how it works. Wireless sensors are installed throughout the plant to collect vast amounts of data in real time. The environment of intense heat and electrical machinery can distort the readings. However, this is compensated by the huge quantity of information generated from both wired and wireless sensors. Increasing the frequency and number of locations of sensor readings serves to offset the distortion. By continuously measuring the stress on pipes, BP learned that some types of crude oil are more corrosive than others—a quality it couldn't spot prior to relying on big data.

5.5 Make Connects the Value Chain

Big data applications have an unbounded potential to generate value for the Make lever of the supply chain—from improvements in product design to labor utilization and capacity planning. Data and analytics applications integrated across the extended enterprise have been shown to raise productivity both by increasing efficiency and improving the quality of products. Benefits exist across all global enterprises. For example, in emerging markets, operational improvements can build a competitive advantage that goes beyond relatively low labor costs. In developed markets, companies can use big data to reduce costs and deliver greater innovation in products and services.

5.5.1 Shared Data

Achieving many of these operational improvements requires access to data from different players in the supply chain. It is not something a company can achieve alone. The challenge is that different members of the supply chain have access to different types of data. IT systems installed along the entire supply chain to monitor the extended enterprise are creating large stores of increasingly complex data. However, this data currently tends to reside only in the IT system *where* it is generated. To use this data for maximum gain, it needs to be shared across the supply chain. This is a major challenge leading companies need to address.

Manufacturing stores more data than any other sector, according to a recent estimate. In fact, the number is close to 2 exabytes of new data stored in 2010.[25] Manufacturing generates data from a multitude of sources, from instrumented production machinery used in process control, to systems that monitor the performance of products that have already been sold. Consider the earlier example of Western Digital, which monitors quality of disks during every aspect of the production process. Similarly, during just a single cross-country flight, a Boeing 737 generates 240 terabytes of data.[26] This amount of data generated will continue to grow exponentially as more sensors are embedded in every aspect of Make. The number of RFID tags sold globally is projected to rise from 12 million in 2011 to 209 billion in 2021.[27] Manufacturers will also begin to combine data from different systems, including, for example, computer-aided design, computer-aided engineering, computer-aided manufacturing, collaborative product development management, and digital manufacturing, and across organizational boundaries in, for instance, end-to-end supply chain data.

Make has operational sensor data. To optimize performance, however, this is not enough.

Sell has customer data that drives supply chains and needs to be combined with operational data to drive decisions. Demand-planning processes, to be done effectively, require customer data from retailers such as POS data. To access such pools of data, manufacturers and suppliers will need to strategically plan the right value propositions and incentives for retailers to be willing to share. Many retailers, for instance, guard customer data as proprietary, but there are many examples of successful data sharing. Walmart is a great example of requiring all suppliers to use its Retail Link platform.

Demand forecasting and supply planning are two important aspects of Make that rely on POS data and other data obtained by the Sell side. Far more value can be unlocked when companies are able to integrate data from other sources. This includes data from retailers that goes well beyond sales. It includes promotion data—such as items, prices, and sales. It includes launch data—such as specific items to be listed and associated ramp-up and ramp-down plans. It also includes inventory data, such as stock levels per warehouse and sales per store. This data is essential for the supply chain to deliver what and when items are needed. Through collaborative supply chain management and planning, companies can mitigate the bullwhip effect and better smooth out flow through the supply chain.

5.5.2 Matching Supply and Demand

Sales & Operations Planning (S&OP) is an important organizational process that cuts across the organization connecting Make with Sell. It is also a highly data-driven process. S&OP is an integrated business management process intended to match supply and demand through functional collaboration. It is a process through which the executive leadership team achieves strategic focus, functional alignment, and synchronization across the entire organization. Done correctly, the process allows executive management to anticipate business changes without resorting to late, reactive, and costly responses. Without this coordinated effort, companies are typically plagued with short planning horizons, lack of functional alignment, and reactive responsiveness.

S&OP cannot happen without data. The process crosses organizational boundaries where internal tensions between departments may create barriers to implementation. The S&OP process relies on having reliable, clean, and accurate data. It requires data to be aggregated from multiple functional organizations. Trustworthy data serves as a basis for the S&OP "conversation."

The benefits of this level of coordination propagate through the entire supply chain. They help companies to use cash more effectively and to deliver a higher level of service. Given the importance of coordination between Make and Sell, best-in-class companies are accelerating the frequency of their planning cycles to synchronize them with production cycles. Indeed, some manufacturers are using near-real-time data to adjust production. Others are collaborating with retailers to shape demand at the store level with time-based discounts. All this coordination is taking place as a result of big data.

6

Impact on "Move"

In 2007, the arrival of *Harry Potter and the Deathly Hollows* was full of heightened anticipation for fans of the series. The 12 million copies of books—a record first printing in publishing—were scheduled to hit the shelves of thousands of retail stores all over the United States in synchronized fashion on July 21 and be available through online bookstores. To minimize the risk of someone leaking the book's ending, strict security was in place. Marketing was pushing a big publicity launch and the mystery of the ending added to the excitement. For the executives at Scholastic, however, this was a major headache. Achieving synchronized deliveries of millions of books at thousands of different locations and meeting deadlines all depended on logistics.[1]

Everyone in the supply chain had to work in tight coordination and was monitored in real time. This included executives from Scholastic's manufacturing and logistics divisions, printers, distributors, and trucking companies, all working to make sure the deliveries, tight schedules, and turnarounds were met. Scholastic bypassed its own warehouses to reduce time, using hired trucks to ship from six printing sites directly to large retailers like Barnes & Noble and Amazon.com as well as numerous distributors of independent booksellers. To expedite loading and unloading, trucking companies such as Yellow Transportation and J. B. Hunt Transport Services used same-size trailers and pallets. To ensure security, every trailer shipping the Potter books had a GPS transponder that would alert Scholastic if the driver of the trailer veered off designated routes.

The timing was especially tricky for online retailers, which had to ship in advance for the books to hit customer doorsteps on July 21. Barnesandnoble.com even developed special data algorithms that enabled its shipping team to figure out when to release books to the U.S. Postal Service (USPS) or United Parcel Service (UPS) to ensure a simultaneous arrival around the country on that day. Who could have imagined that printing and delivering books could be such a digital nightmare?

6.1 Moving the Things We Sell

Move is essentially logistics. Logistics is the business function responsible for transporting and delivering products to the right place at the right time throughout the supply chain. It plays a critical role in the functioning of both the organization and the supply chain. Without logistics, materials would not arrive when and where they are needed. To achieve this, logistics must plan and coordinate all material flows from source of supply to all users in the chain. Even the delivery of Harry Potter books is much more complicated than it may appear (see Figure 6.1).

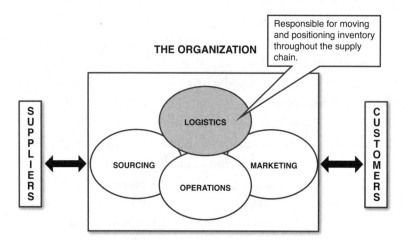

Figure 6.1 The logistics function

6.1.1 Delivering the Perfect Order

Logistics, however, is about much more than just movement and storage of product inventories throughout the supply chain. Logistics connects the organization to both its customers and suppliers. It works with sourcing to link the organization to its suppliers. It ensures that materials are delivered inbound when needed. Logistics works with operations so that sourced materials are transformed into finished products. It also works with marketing to ensure distribution and delivery of outbound products to external customers. Logistics is about delivering what is known as "the perfect order" to customers.

The basis for the term *the perfect order* is what is known as the "five rights" of customers. These are receiving the *right product* at the *right place* at the *right time* in the *right condition* for the *right cost*.[2] The definition of the perfect order is one that meets all of the following conditions of delivery:

- **Complete**—All items are included in the order in the quantity requested.
- **On time**—The order arrives on the customer's request date, using the customer's definition of on-time delivery.
- **Accurate**—All documentation is included, such as packing slips and invoices.
- **Perfect condition**—The order is in the correct configuration, is damage free, and is faultlessly installed.[3]

Achieving the perfect order requires optimization and coordination of all activities of the supply chain. As the pace of commerce has increased, so have customer expectations. "Better, faster, and cheaper" is no longer good enough. Today's customers are demanding perfect orders, shipped and delivered on time to the minute, at a cost that leaves little margin. It is up to logistics to make this happen.

6.1.2 Coordination

Moving goods requires coordination. Without this, there is no perfect order. This includes managing the entire distribution network, location of warehouses, distribution centers and plants, and coordinating the modes of transportation between them. It also includes design and management of operations throughout the network for efficient storage and quick movement of goods. Recall that in the Harry Potter example, pallets of the same size were used to expedite movement, and consider that even seemingly small decisions such as size of trailers can be significant.

So what does logistics have to do to coordinate movement of goods? Here are some activities:

- Customer service
- Demand planning and forecasting
- Inventory management
- Materials handling
- Order processing
- Packaging
- Reverse logistics
- Transportation
- Warehousing

Big data analytics has had a tremendous impact on improving the activities involved in Move, such as the ability to assess and adjust actual logistics performance in real time. This has included the ability to monitor customer requirements, production demands, and inventory levels throughout the supply chain and to respond in real time to these requirements. Analytics and IT are the lifeblood of this coordination. Such systems must be integrated internally to take into account marketing and production activities. They must also be integrated with others in the supply chain, to provide accurate information throughout the channel, from the earliest supplier through the ultimate customer. This is what leading companies do.

Recall Walmart's coordination capability through its Retail Link used to communicate with its suppliers. Information flows from Walmart to suppliers and back to Walmart in real time. Walmart uses bar-code readers in its retail store checkout lanes for capturing real-time sales information. This is then downloaded to suppliers in real time. Suppliers use this information to determine the products they need to ship to Walmart. There is no lag time because orders are created automatically as they are needed. RFID technology keeps track of inventory quantities and their locations, so orders can be optimized. The information also flows the other way. Walmart receives information from suppliers relating to shipment status, delivery schedules, quantities, and even billing/invoicing. This system provides suppliers with rapid feedback on sales so that they can anticipate production requirements based on accurate, near-real-time sales data. They also receive payments earlier, improving cash flow. What are the benefits? Walmart no longer has to place orders directly with many suppliers and can keep inventory levels to a minimum. This coordination is made possible through technology, and it reduces inventory costs and improves customer services (see Figure 6.2).

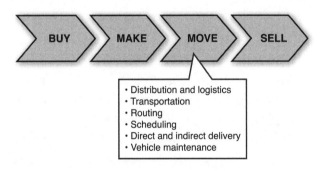

Figure 6.2 Big data applications in logistics

6.1.3 It Isn't Cheap

Logistics is costly. It is a function that requires large investments in infrastructure, such as transportation vehicles, material-handling equipment, and information technology. For this reason, many companies outsource this function to outside firms. Companies such as UPS, Federal Express (FedEx), and DHL have made it their business to provide logistics services to their clients. These companies have invested in the needed infrastructure, such as their own fleet of airplanes and trucks and the latest technology to track packages from point of origin to destination using bar-code technology and scanners. In addition to moving products, these companies can be contracted to perform a range of logistics functions, including designing an entire distribution network for their clients. They can also provide services such as management of inventory, warehouse management, and even customer interface.

Peter Drucker underscored the cost of logistics:

"Almost 50 cents of each dollar the American spends for goods goes for activities that occur after the goods are made, that is, after they have come in finished form...distribution is the process in which physical properties of matter are converted into economic value; it brings the customer to the product."[4]

We can argue the 50-cent estimate. However, what is not debatable is the high cost involved in Move. Control of costs is one of top management's most significant concerns in the new millennium. As a result, efficient and effective control of logistics is an imperative. One aspect that adds substantial cost to logistics is the need for technology—information and processing technology, both of which rely on data. Achieving efficiency and scale needed cannot be done without it.

6.1.4 Technology

Logistics is a data and technology-enabled function. Logistics must also have access to information in real time to be able to track product movement and plan exact timing and location of deliveries.

It must ensure security of goods from theft or tampering while the products are moved. Recall in the case of Harry Potter that all trailers carrying books had GPS transponders to allow the company to monitor product location in real time.

Today, organizations are utilizing various technologies to reduce logistics costs and improve customer service at the operational level, as well as implementing technology initiatives related to data protection, network security, disaster recovery, data warehousing, and data analytics.[5] All this is dependent upon big data analytics to drive and coordinate these systems. And it is not cheap.

Traditional methods of managing logistics activities are simply insufficient in today's fast-paced economy. To achieve the needed efficiency—speed, accuracy, scale, at low cost—implementation of the latest technologies is necessary. There has been a proliferation of technological developments in areas that support logistics. And they all use data to one extent or another.

They include artificial intelligence (AI), electronic data interchange (EDI), bar-code scanning, local area networks (LANs), point-of-sale (POS) data, radio-frequency identification (RFID), satellite data transmission, smartphones, and tablet PCs. Many manufacturing and merchandising firms have employed these newer technologies to reap financial and customer service benefits. In fact, more than one third of manufacturers will increase their use of smartphones and tablet PCs in the near future.[6]

Software applications will continue to be important as well, such as warehouse management systems (WMSs), transportation management systems (TMSs), inventory optimization, Enterprise Resource Planning (ERP), supply chain planning (SCP), global trade management (GTM) software, and labor management systems. Each of these technologies provides an independent source of data. Connecting these data sets has the potential to revolutionize supply chain management.

Big data analytics applications are increasingly becoming familiar to many logistics companies. However, the investments to use them are often too large. Adoption needs should focus on strategic drivers as not every capability is equally important. One alternative is to jointly determine and analyze needs with other supply chain partners, given that logistics touches all entities along the supply chain. There may be opportunities to develop joint platforms and share in costs (think Walmart's Retail Link) or strategically evaluate the degree of technology needed (do we need RFID tags for all items?).

6.1.5 Analytics on the Move

Analytics has been employed along the Move lever for quite some time. Analytics has been used to optimize inventory, determine product quantities, optimize locations, identify optimal distribution center locations, minimize transportation costs, and even analyze vehicle capacity and cost for each trip. These approaches have been developed under the term *operations research* and were among the earliest analytical approaches in business. Leading-edge companies—especially third-party logistics providers, such as UPS, FedEx, and DHL—have been using analytical models to optimize transportation, routing, and maintenance schedules.

Analytics applications along the Move lever, however, do much more than just manage inventory movements across global supply networks. Companies are using analytical applications to design supply chains that are more flexible, optimized, and event driven. Such designs enable rapid resupply of fast-moving goods and minimize stock-outs. These supply chains are more flexible and can integrate multiple channels. This is especially important as retailers expand their offerings through multichannel distribution.

Amazon is a leader in big data applications for the Move lever. The company has built a new supply chain process and systems that generate accurate forecasts at the SKU level and support fulfillment, sourcing, capacity, and inventory decisions.[7] These algorithms mine and coordinate data based on historical demand and event history, and they generate forecasts for each fulfillment center, inventory planning, procurement cycles, and purchase orders. Similarly, Office-Max has put in place new systems to optimize inventory at the SKU level, transportation cost, and warehouse investment. The company analyzes store product movements to drive both product assortments and restocking. Other variables considered are store promotions and effectiveness of promotion execution. German department store retailer Metro Group uses RFID to detect the movement of goods within stores and records patterns of movement on and off the shelf for later analysis.[8] This data is placed in dashboard format, providing store managers with information on onsite inventory and giving automated out-of-stock alerts.

One of the primary drivers of Move is the availability of extensive data—in particular, sensor data. New sensors, such as RFID, are making dramatic amounts of data increasingly available for the next generation of supply chains. Some companies have been using RFID analytics for years. For example, Daisy Brand, a dairy products manufacturer in the United States, began using RFID analytics in 2005.[9] The company uses it to track how long it takes products to reach the store shelf as well as to predict replenishment rates. The company has found this particularly useful during promotional periods. In addition to RFID data, Daisy Brand also makes extensive use of data obtained through Walmart's Retail Link data. The company obtains weekly point-of-sale and inventory information on which it conducts extensive analysis.[10]

Companies are using analytics well beyond management of inventories along the Move lever. Companies such as UPS use analytics for vehicle preventive maintenance, in order to prevent transportation failure.[11] UPS operates more than 96,000 vehicles. Vehicle breakdown for any reason can disrupt services, delaying time windows for deliveries and pickups. UPS has used predictive analytics since the late 2000s to monitor its fleet of vehicles in the United States. As a result, UPS knows when to perform preventive maintenance. A breakdown on the road can cause major disruptions and UPS used to replace certain parts after two or three years, even when the parts were fine. Since switching to predictive analytics, the company has saved millions of dollars by measuring and monitoring individual parts and replacing them only when necessary.

6.2 How Big Data Impacts Move

Big data analytics comes to bear on logistics through customer service functions, transportation of goods, warehousing, location optimization, inventory management, reverse logistics, and energy use and sustainability performance. We discuss these areas in turn.

6.2.1 Customer Service

Customer service is the ultimate goal of the Move lever and involves delivering the perfect order. It is achieved by all the logistics activities working in tandem. Customer service can be defined as "a customer-oriented philosophy that integrates and manages all elements of the customer interface within a predetermined optimum cost-service mix."[12]

Today's consumers expect the perfect order. At a minimum, the perfect order means shipping all items ordered (measured as "percent order complete"), without error (measured as "percent order accuracy"), and on time to the customer (measured as "percent on-time delivery"). This raises an important question. What does the customer really mean by "on time?" The recent trend has been moving toward a standard of a minimum of next-day shipping. However, an increasing number of retailers are moving toward offering same-day shipping, and a few are even offering same-hour shipping for selected SKUs in selected geographic areas.

Here is the problem. Such high levels of service are expensive. Retailers that have a multichannel distribution network can possibly provide same-hour shipping and delivery from a brick-and-mortar store. However, this is only feasible within a certain geographic area. And it is still costly. It is for this reason companies should ask themselves some important questions:

- Do all our customers really require next-day shipping?
- Is this necessary to achieve high customer service?

For some market segments, this will certainly be true. Many customers, however, may prefer a discount on the shipping price in exchange for later delivery. Offering these options provides distribution operations with flexibility that can help with smoothing order demand and balancing staffing. Also, relaxing next-day service levels requirements—even to 48-hour delivery—significantly reduces the capital investment needed as flexible labor can be used for added capacity.

Analytics is the solution to identifying which segments respond to which elements of customer service. Microsegmentation—using software applications and analysis—reveals exactly the customer service requirements that are order winners in each segment. It can create segments of customers for which such high delivery standards are critical. It then permits allocating distribution costs to those segments—rather than averaging it across segments and including those that are indifferent to such a costly proposition. This then provides a better sense of true costs.

6.2.2 Transportation

Transportation is a major logistics activity. It involves the movement and flow of goods from point-of-origin to point-of-consumption. Transportation is often the single largest cost in the logistics process. Therefore, it is a critical component that must be managed effectively.

Transportation involves decisions such as selecting the method of shipment—air, rail, water, pipeline, truck, or intermodal. The most common is the latter, which involves selecting a mix of mode options. Air is fastest but is expensive and has capacity constraints; water and rail are cheapest but are slow; pipeline has geographic limitations. Transportation decisions also involve choosing the specific path (routing); complying with various local, state, and federal transportation regulations; and being aware of both domestic and international shipping requirements. Big data applications can optimize the mix considering multiple criteria, including delivery times, speed, capacity, security, cost, as well as other factors such as minimizing the carbon footprint.

6.2.2a Route Optimization

One big data application in transportation has to do with route optimization—selecting the best route for a select criterion. Smart routing is based on digital maps and real-time traffic information, and is one of the most heavily used applications. The applications optimize transportation by using GPS-enabled big data telematics—the remote reporting of position—to determine the best route to travel. The application can also be used to optimize fuel efficiency, preventive maintenance, driver behavior, and vehicle routing. It can also analyze transport routes from source to destination. This capability can substantially improve transportation costs, efficiency, as well as fleet and distribution management.

The more advanced navigation systems can receive information about traffic in real time, including accidents, scheduled roadwork, and congested areas. These systems are also capable of giving users up-to-date information on points of interest and impending weather conditions. Some of these devices can not only provide drivers with recommendations on which routes to take to avoid congestion, but also report back information on location and movement to a central server. This allows congestion to be measured even more accurately and creates a feedback loop that in turn provides better navigation.

6.2.2b Automotive Telematics

Most major transportation carriers today deploy Global Positioning System (GPS)–based telematics devices in trucks and trains. These devices provide a wide variety of data about driving behavior, speeds under various conditions, traffic, and fuel consumption. Companies such as UPS and Schneider Logistics have already employed telematics data to redesign logistical networks. UPS, in fact, is using telematics data to redesign and optimize its entire delivery network for only the third time in its more than 100-year history.

These types of systems also have other sensing and monitoring capabilities. For example, they can alert drivers to when vehicles need repairs or software upgrades. They can also locate vehicles during emergencies, such as when air bags have been deployed.

You saw how UPS uses data for preventive maintenance. However, UPS uses its geo-loco data for many other aspects of logistics—from route optimization to labor management.[13] Its vehicles are fitted with sensors, wireless modules, and GPS, all gathering data. This data is then fed into headquarters, which uses algorithms to predict more than just engine trouble and schedule maintenance. It also lets the company track routes in real time. It knows the location of any vehicle, which can be particularly useful during delays. The company can also use it to monitor drivers and scrutinize their itineraries to optimize routes. The system also improves safety and efficiency. For example, the algorithm identifies routes with fewer turns that must cross traffic intersections.

6.2.3 Warehousing

Traditionally, the role of warehouses has been to provide storage space for goods, as well as inbound and outbound transport. Goods arrived at the warehouse via truck or rail, were unloaded, and then were placed in storage bins in the warehouse. When customer orders arrived, goods were picked up from their storage location—known as *order picking*—staged for transport, loaded on to trucks or rail, and shipped.

Contemporary warehouses are increasingly places for mixing inventory assortments to meet customer needs. They are about enhancing and expediting movement rather than lengthy storage. In fact, lengthy storage is discouraged as it does not add value and just contributes to cost and obsolescence. Warehouses provide a centralized location that stores and organizes inventories of products before they get distributed to customers. For that reason, they are often called distribution centers.

Warehouses or distribution centers play a critical role in the supply chain, enabling efficient movement of goods on both the inbound and outbound side of the organization. For example, in just-in-time (JIT) and lean manufacturing, warehouses can be located close to the manufacturing facility, enabling frequent deliveries of materials on a just-in-time basis. At the same time, warehouses can be utilized to create product assortments for customer shipments. A strategically located centralized warehouse, for example, can take advantage of consolidated shipments. The products can then be sorted and arranged for a particular customer and then shipped. In addition to these benefits, many warehouses are increasingly performing tasks traditionally done at a manufacturing or retail site. This may include repairing items, putting garments on hangers and sequencing them to be rolled straight on to the retail floor, and adding labels and price tags. Optimizing these decisions is a role for big data.

6.2.3a Labor Utilization

Labor is the greatest expense item associated with warehousing and tasks such as picking and packing individual orders. This is especially problematic in a direct-to-consumer fulfillment, generating the largest share of those labor costs. One of the most time-consuming aspects of labor is travel. Warehouses are large facilities. Associates walking or riding around the building en route to and from product storage locations is one of the most wasteful activities. Analytics can be used to create material-handling solutions that reduce travel, such as developing picking assignments that minimize distance walked. It can create pick modules and zone assignments that meet order requirements while optimizing order picker access to SKUs. All this serves to substantially improve efficiency of labor and warehouse operations.

6.2.3b Material Handling

Another warehouse challenge is material handling. Material handling is concerned with every aspect of movement of goods— including raw materials, in-process inventory, and finished goods within a plant or warehouse. The objective of material handling is to eliminate handling wherever possible, minimize travel distance, and minimize goods-in-process. In addition, efficient material handling should provide uniform flow that is free of bottlenecks and should minimize losses from waste, breakage, spoilage, and theft.

Costs are incurred every time an item is handled. Because handling generally adds no value to a product, handling should be kept to a minimum. For items with low unit value, the proportion of material-handling costs to total product cost can be significant. Analytics applications can optimize this flow. By carefully analyzing material flows and planning efficient routes, these applications can save the organization significant amounts of money.

6.2.3c Analytics Solutions

Today's warehouse management systems (WMSs) are increasingly sophisticated. This is in no small part due to data analytics applications that make up the solutions. These software applications are designed to provide management with the information needed to efficiently control movement of materials within a warehouse and support the day-to-day operations. WMS programs enable centralized management of tasks such as tracking inventory levels and stock locations and can be either stand-alone applications or part of an Enterprise Resource Planning (ERP) system.

Early warehouse management systems could only provide simple storage location functionality. Current WMS applications can be so complex and data intensive that they require a dedicated staff to run them. High-end systems may include tracking and routing technologies such as radio-frequency identification (RFID) and voice recognition.[14] Many systems enable coordination of inventory and sales across multiple global warehouse systems. These capabilities enable tracking of global inventory from one screen where sales staff can enter orders or monitor performance from their iPhones.

Another data-driven aspect that has changed warehouse operations is automated material-handling systems.[15] These systems enable data-driven automation that improves accuracy and efficiency and easily adjusts capacity requirements. More sophisticated systems automatically slot products based on their velocity, with fast-moving products stored near workers. As a SKU's velocity changes, its storage location also changes automatically. Some also adapt to workers' individual abilities with high-value items being sent to long-term workers, for example. The grocery store chain Kroger was the trailblazer when the company invested in the first automated full-case picking system in 2004 in its distribution center in Arizona.[16] This was so successful that the company quickly expanded the system to other locations. More recently, Target Corporation implemented a new, fully automated, 360,000-square-foot perishables and frozen food distribution center in Texas.[17] These large, automated, data-driven warehouses offer unprecedented efficiency and capability. Kiva Solutions, for example, is a company that offers automated material handling where operators stay in work stations and the system delivers any product to any worker at any time, resulting in lower labor costs and shorter cycle times.[18] The optimized system achieves incredibly high picking accuracy through innovative station-based pick-to-light, put-to-light and scanning capabilities, which are available for all items.

6.2.4 Location Optimization

Facility location is determining the best geographic location for a company's facility. Facility location decisions are one of the most important areas of operations management. The reason is that these decisions require long-term commitments in buildings and facilities, which means that mistakes can be difficult to correct. Second, these decisions require sizable financial investment and can have a large impact on operating costs and revenues. Poor location can result in high transportation costs, inadequate supplies of raw materials and labor, loss of competitive advantage, and financial loss. For these reasons, businesses have had to think long and hard about where to locate a new facility.

The process has historically involved geographic and demographic data analysis. Many relatively crude mathematical approaches have been applied in the past, such as factor analysis and the *load-distance model*—a procedure for evaluating location alternatives based on distance. Big data now uses analytics to optimize location sites and formats. These procedures are quickly becoming more sophisticated as they apply more advanced analytics, such as probabilistic algorithms, and leverage ever-more detailed and varied data sources.

There are two aspects that are changing location-based decisions. First, today's quality and granularity of data can be used to make much more detailed location decisions. In the past, companies based analyses primarily on census data about populations and incomes in target geographic areas. This information was often dated. Today's data is more current—updated frequently versus every decade. Today's data is also more varied. It includes both demographic data and also psychographic data. This is information on personality types, values, attributes, interests, and lifestyles. The data also includes information on competitor stores, customer data from loyalty programs, and shopper intercepts. This data can also be used for *what-if* analysis of various performance drivers. The second aspect that is different is that the new applications allow simultaneous decisions with regard to optimizing both the location and format of the site—even whether to remodel or what merchandising approach to use for that market.[19]

Consider the site selection approach used by office retailer OfficeMax.[20] The company has more than 900 sites and the new process has allowed it to double its pace of new store openings. What is different? Rather than the usual focus on demographics, the company has prioritized its core markets in terms of economic potential. The system takes into consideration a number of factors. It includes customer information to consider the potential of cannibalization. It also considers the proximity to existing distribution networks as one factor in the economics of a new store. Other data includes marketing, finance, and store operations. Thus far, the new stores have produced results in line with the forecasting model's predictions.

6.2.5 Inventory Management

Inventory management involves ensuring that the right quantities of materials are in place when and where they are needed. Inventories are also costly. Inventory holding cost, for example, can in many cases be 50 percent of cost of goods. Maintaining raw materials, parts, work-in-process, and finished goods inventories consume physical space, personnel time, and capital. Money tied up in inventory is not available for use elsewhere. On the other hand, reducing holding costs means a reduction in inventory levels and results in frequent replenishment and possibly stock-outs. This is the classic struggle in inventory management.

Inventory management involves trading off the level of inventory held with frequency of orders. The goal is to achieve high customer service levels, while minimizing the cost of holding inventory, including capital tied up in inventory, warehousing costs, and obsolescence. Successful inventory control involves determining the level of inventory necessary to achieve the desired level of customer service while considering the cost of performing other logistics activities.

Sensors are changing the way we manage inventory. RFID tags and readers are increasingly providing substantially more data on product movements and locations for retailers to analyze. RFID and telematics sensors primarily track location. However, the more recent ILC (identification, location, condition) sensors can also monitor the condition of goods in the supply chain.[21] ILC sensors monitor such variables as light, temperature, tilt angle, gravitational forces, and whether a package has been opened. They can transfer data in real time via cellular networks. Obviously, the potential to identify supply chain problems in real time and take immediate corrective action is greatly enhanced with this technology. We have only begun to consider how analytics might be used to enhance the value of ILC-generated data.

6.2.6 Reverse Logistics

Reverse logistics is the process of moving products upstream from the customer back toward manufacturers and suppliers. This occurs for a variety of reasons, such as damaged products, warranty repairs, replacement, remanufacturing, product recalls, recycling, or items the customer simply did not want. The key issue here is that reverse logistics is expensive. In fact, the cost of moving a product back through the supply chain from the consumer to producer may be as much as nine times the cost of moving the same product from producer to consumer. According to the Reverse Logistics Association, the annual volume of returns is somewhere between USD $150 and $200 billion. The costs associated with all of those returns average between 7 and 10 percent of the cost of those goods returned.[22]

In addition to costs of material handling and transportation, there are other issues that add complexity to reverse logistics. One is handling the financials and the cash flows once items are returned. Another is arranging for warehouse and storage space in the reverse order that does not confuse or take away from flow that is occurring in the downstream direction. Yet another is abiding by "green laws" in countries that have them that may require returning packaging materials for proper disposal.

Just as big data analytics is impacting the downward flow of products, it is also having an impact on reverse logistics. There are many reasons for the return of goods. Analytics can help optimize store layouts and locations, as well as product design and packaging to help minimize returns. Route analysis can optimize the material-handling needs and proper management of backward flow inventory. Further, scanners—such as RFID—can monitor inventory movement of product returns.

The company Inmar, for example, provides real-time supply chain data and analytics for products as they move through the reverse supply chain.[23] The company provides analysis solutions that help companies in monitoring store activity and developing programs that help minimize product returns and improve their supply chain.

Inmar engineers and analysts collect and analyze large quantities of data to deliver studies about the most detailed inner workings of the customer's supply chain and those of its trading partners. These consulting studies are then used by the customer to implement packaging changes, improve distribution processes, and assess the effectiveness of existing buying and selling practices.

6.2.7 Energy and Sustainability

Most companies recognize sustainability as important to their mission.[24] Companies like Walmart, Unilever, General Electric, and Marks & Spencer are examples of companies that are considered sustainability leaders. In a 2011 research study, Unilever was identified as a leader in sustainability.[25] The company initiated its *Sustainability Living Plan,* which has as its goals for 2020 to improve the health of one billion people, purchase 100 percent of its agricultural materials from sustainable sources, and reduce by one half the environmental impact of its products. At the same time, the company anticipates doubling its revenues.

What does this have to do with big data applications in logistics?

The Move lever—logistics—is extremely energy intensive. It is one of the most significant places to reduce energy consumption. Also, big data analytics permits the development of optimized solutions for countless problems that take into account numerous variables. Adding sustainability concerns to the ones discussed thus far is a minor challenge for these applications.

From an energy perspective, organizations have made significant strides in minimizing their energy consumption. Given the higher cost of petroleum and petroleum-based fuels, there have been financial benefits to go along with the environmental benefits. In fact, energy consumption in the U.S. manufacturing sector declined by 10 percent from 2006 to 2010.[26] This documents the strides that companies have made in energy conservation.

Virtually all the activities of logistics can use big data analytics to include improvements in sustainability metrics, while still maintaining all the other objectives. Consider facility location. In addition to optimizing market coverage, location optimization can include placing facilities closer to markets to reduce the requirement for long transportation to reduce energy and emissions. It can involve selection of warehouses and distribution centers to reduce inventory and transportation cost; it can include constraints such as location of facilities in developing countries, which can give rise to social issues of fair wage rates, child labor, and compliance with local environmental regulations.

Big data analytics can be used to optimize metrics such as food miles, how much energy it takes to deliver to final consumer, effects of truck traffic, comparison of centralized versus decentralized warehousing on long-term energy costs, and cumulative impacts of fuel. Warehouse designs can include consideration of the inclusion of energy-intensive storage facilities, Leadership in Energy & Environmental Design (LEED) certification, energy efficiency, graywater recycling, efficient light bulbs, powering down when not in use to conserve energy, and new building designs. Although these efforts appear to cost more, that may not be the case when amortized with lower overhead, consideration of more efficient heating and cooling, and tax incentives. Analytics can help managers create a business case to justify implementation of these energy-saving measures.

The Rodon Group is an example of a company committed to environmental sustainability.[27] The company's green initiatives make it a leader in the plastics industry. Its manufacturing process minimizes packaging, transportation costs, waste, water consumption, and energy. The company also relies on extensive data gathering and measurement to implement these sustainability efforts.

6.3 Integrating Logistics Activities

Move—or logistics—is a system of many intertwined activities. They must be coordinated, managed, and optimized together. Companies that do not adopt a systems approach to logistics find that it becomes a fragmented and uncoordinated set of activities. The various activities end up being spread throughout various organizational functions with each individual function having its own budget and set of priorities and measurement. These include activities like the ones discussed previously—such as customer service, transportation, warehousing, inventory management, order processing, and production planning. This results in poor performance and higher costs. Leading firms—such as 3M, Dell, Quaker Oats, and Whirlpool Corporation—have found that logistics costs can be reduced by integrating logistics. Without this integrated approach, inefficiencies result. This results in excess inventory, poor customer service, and a very common occurrence—too much inventory in the wrong place and not enough where it is needed. This is a significant task and it cannot be accomplished without information technology and today's big data analytics applications.

7

Impact on "Buy"

IBM exemplifies an analytics-driven company. A decade ago, IBM used analytics to move from a regional and globally dispersed procurement to a single integrated global supply chain.[1] Today, the tech giant continues to use analytics to manage every aspect of its supply chain—from cost containment and risk management to using software to model the impact of potential scenarios on supplier networks. Its latest supply chain initiative involves using *predictive and prescriptive analytics* to drive operational improvements.

Predictive analytics helped IBM respond in a timely and effective way when natural disasters threatened to disrupt the company's supply chain. In April of 2010, a volcano in Iceland halted flights throughout much of Europe. The analytical software told IBM to focus on Asia rather than Europe, as a response. The software indicated that the critical link in IBM's supply chain was Hong Kong. Management was curious. This did not make sense. Then it quickly became clear. The software was correct and much smarter.

The analysis forecast that if IBM did not take steps to secure sufficient airlift once the volcanic eruption abated and flights resumed, it would encounter a bottleneck in Hong Kong. This would occur as a result of the company quickly moving a backlog of components and products from Asian manufacturers to European customers. IBM listened. As a result of that prescriptive analysis, the company booked space on commercial and charter aircraft from Hong Kong to Europe in plenty of time. "We didn't sit and watch what was going on with the disaster," said Timothy E. Carroll, vice president of supply chain operations. "We prepared ourselves for what to do once the disaster lifted."[2] Now, a dedicated research arm within IBM's supply chain organization reviews various scenarios to prepare a response to a natural disaster or man-made crisis anywhere in the world.

7.1 Big Data and Buy

The Buy lever of the supply chain is the sourcing function. Sourcing is responsible for all activities and processes required to purchase goods and services (see Figure 7.1). The sourcing function has responsibility for the supply side of the organization addressing the upstream part of the supply chain. Consider that the supply area makes up between 50 and 85 percent of revenue for most manufacturing organizations. In fact, more than 80 percent of cost of the typical automobile accounts for purchased items.[3] Therefore, the Buy lever needs to be managed carefully.

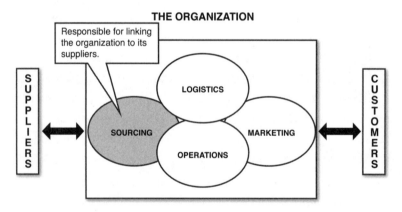

Figure 7.1 The sourcing function

7.1.1 The Scope of Buy Decisions

The challenge in managing sourcing is that it is complex. The sourcing function is often referred to by names such as *purchasing*, *procurement, sourcing, strategic sourcing*, and *supply management.* Although these terms are often used interchangeably, they do not necessarily mean the same thing. These terms also underscore the breadth of decisions for which sourcing is responsible. The big data analytics applications selected along this lever need to consider the differences in focus of what the company needs.

Purchasing is a term that defines the process of buying goods and services. It is a narrow functional activity with duties that include supplier identification and selection, buying, negotiating contracts, and measuring supplier performance. However, it is much more than buying. Over the years, the purchasing function has evolved to include a much broader and more strategic responsibility, which is termed *strategic sourcing* or *supply management.*

Strategic sourcing involves a much more progressive approach to the sourcing function. It goes beyond focusing on just the price of goods to looking at the sourcing function from a strategic and future-oriented perspective. It considers sourcing opportunities that will solve greater problems for the buying firm and give it a competitive advantage. This requires expanding the role of sourcing from mere buying to building close and longer-term working relationships with specially selected suppliers and partners (see Figure 7.2).

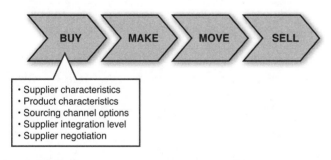

Figure 7.2 Big data applications in sourcing

7.1.2 How Big Data Impacts Buy

Selection of big data applications along the Buy lever depends in part on whether you are improving purchasing or managing strategic sourcing. Big data offers a range of possibilities from improving current processes of containing spend to enabling advanced risk scenario planning. Regardless of scope, it is the convergence of data from different sources and internal systems that can result in making considerably faster and better sourcing decisions.[4]

7.1.2a Order Processing

One of the most basic applications of big data analytics that leads to quick improvement is in order processing. Order processing is the cycle from order placement to order fulfillment. The speed and accuracy of a firm's order-processing activities have a great deal to do with the level of customer service the company provides. The order-processing cycle is the customer interface with the organization and how this is handled impacts customer perceptions of service and satisfaction. Big data analytics can have a huge impact as it enables efficient and effective management of the order-processing cycle. IT systems, in turn, provide control and visibility over the entire life cycle of a transaction. This ranges from the way an item is ordered to the manner in which the final invoice is processed.

IT and big data analytics can help to reduce the time between order placement and product shipment. Orders can be transmitted in real time from buyers to sellers via the Web or electronic data interchange (EDI). Automatizing this system substantially improves both order-processing accuracy and response time. Also, automatizing often results in additional savings on other logistics expenses, such as inventory, transportation, and warehousing, or increased sales from improved customer service. A thorough financial analysis justifies the cost of the computerized order-processing system and business analytics at this stage.

IBM, for example, uses analytics to manage two critical order-process cycles.[5] The first is the order-to-cash cycle. That process starts when a customer is ready to do business with IBM. It continues with the placement and then the execution of the order, including manufacturing and delivery. The cycle also encompasses billing and invoicing, accounts receivable, and post-sales support. The second cycle is called procure-to-pay, which encompasses purchasing and payment of suppliers. The procure-to-pay systems enable the integration of the purchasing department with the accounts payable department. "It's everything that we do with external suppliers," says Timothy E. Carroll, vice president of supply chain operations. "Our chief procurement officer and his organization have full responsibility for all purchasing on behalf of IBM, whether it's production, administrative, travel, you name it."

7.1.2b Standardization

Automating the order-processing cycle enables standardization and consistency in managing supply. For IBM, process standardization ensures managing a global order cycle and being able to offer the same level of fulfillment regardless of location. Consider a client in Europe who discovers a need for a critical part late at night. The client doesn't have to wait until normal business hours to place an order. Instead, he or she can contact a fulfillment center in another part of the world to process the request. Analytics has enabled coordination and standardization of fulfillment that is supported 24/7 around the globe.

7.1.2c Visualization

Big data analytics has a huge impact on creating new types of applications that help manage spend, as well as improving the ability to use existing applications. The reason is that big data analytics applications enable visualization of large data sets and relationships of variables. This enables presenting information in a way that managers can use it effectively, through the use of images, diagrams, dashboards, or animations.[6] These include the development of interactive briefing dashboards on suppliers or commodities that leverage mash-ups of continuously updated internal and external information.[7]

Managers can now visualize the numbers they want to access through dashboards. These data sets can include aggregated and disaggregated data, from traditional invoice data to contract information per part at the SKU level. They can include information on parts and details of the bill of material (BOM), as well as warranty and claims data. Tremendous insights can be attained when this information is further combined with data from other sources—such as market information, demand data, and performance data. The interactive dashboards can take all this data and allow the user to conduct what-if analysis, allowing the user to see the results in a visualized format.

7.1.2d More and Faster Queries

These new applications also create the ability to run more queries faster. This means running a greater number of scenarios to optimize sourcing award decisions based on internal constraints and different market forecasts. It means the ability to instantly identify optimal points on a massive spend cube. It also enables the ability to look at spend in new ways—by different locations, taxonomies, or material codes. A company can now quickly create new spend cubes as new opportunities come along, then test different scenarios. Big data applications provide speed of queries and ability to respond quickly. This is increasingly becoming the norm as more companies acquire these capabilities.

Big data analytics enables solving complicated sourcing challenges with an eye on total cost. The ability to leverage optimization capabilities from basic e-sourcing suites is becoming an expectation. This enables doing quick queries of large-scale network optimization for common sourcing events, such as Less-Than-Truckload (LTL) or full truckload spend. In leading procurement organizations, the ability to query consequences of infrequent events is also becoming routine. The ability to quickly come up with answers to sourcing questions is rapidly becoming a competitive advantage.

7.1.2e Cost Savings

These big data applications enable companies to make far-better sourcing decisions. In addition to having an eye toward reducing overall total cost, these applications enable optimizing every aspect of the sourcing decision well before the sourcing event. For example, applications now enable collaborations between members of the design team, such as suppliers, engineers, and members of the procurement team. As a result, they can suggest changes to specifications and tolerances resulting in substantial cost savings.

Electronics retailer Best Buy collaborates with all its suppliers.[8] It uses CPFR to exchange reports and analyses with major suppliers. This has resulted in improvements in forecast accuracy and product availability to customers. Best Buy shares information, such as base and promotional forecasts, inventory available and required, and forecast accuracy reports for several weeks out. This is used for joint forecasting and allocation decisions through CPFR.

7.1.2f Predictive Edge

Big data analytics—especially predictive analytics—enables explaining what was recently the unexplainable. The unexplainable may take the form of a rise or decrease in the prices of supply. Or it may be rapidly deteriorating quality or service-level metrics at suppliers that previously were in the top quartile of performance. It may also be sudden scarcity of supply of a previously available material. Gaining a predictive edge on the market and suppliers requires understanding not only *what* areas correlate but also *when* correlations break down. Big data applications add scenario planning, predictive modeling, and forecasting competencies into procurement.

Predictive analytics and scenario planning are some of the biggest big data tools for leading-edge companies. IBM, for example, uses them as a regular part of its toolbox.[9] These applications have enabled the company to foresee problems and take preemptive actions to prevent supply chain interruptions anywhere in the world.

7.1.2g Enables Co-Creation

Integrating and sharing sets of big data, applications, and platforms can bring manufacturers and suppliers together in a process of product *co-creation*. Think about industries such as automotive and aerospace. Here, a new product is often assembled with hundreds of thousands of components supplied by hundreds of suppliers from around the world. Sharing big data across the supply chain enables co-creating designs with suppliers. This can significantly enable experimentation at the design stage. Designers and manufacturing engineers can share data and quickly and cheaply create simulations to test different designs, the choice of parts and suppliers, and the associated manufacturing costs. This is an extremely time- and cost-saving capability as decisions made in the design stage typically drive 80 percent of manufacturing costs. According to a recent estimate, companies such as Toyota, Fiat, and Nissan have all cut new-model development time by 30 to 50 percent as they engage in co-creation with their suppliers.[10] In fact, Toyota claims to have eliminated 80 percent of defects prior to building the first physical prototype.[11]

7.2 How Much Do You Need?

How much to invest in big data analytics on the Buy lever depends on a host of variables, including supply chain complexity, the number of sourcing options, as well as the location of suppliers.

7.2.1 It Depends. How Complex Is Your Supply?

There are countless applications of big data analytics on the Buy lever, ranging from automating the order-processing cycle to segmenting suppliers and informing supplier negotiation. How deep do companies need to go?

Types of complexities drive how to best utilize big data analytics on the supply side. Just as there are uncertainties on the demand side, there are also uncertainties on the supply side. The supply side of the supply chain can be classified as a *stable* or an *evolving supply process*.[12] A stable supply process is where sources of supply are well established, manufacturing processes used are mature, and the underlying technology is stable. Consider sourcing apples, lumber, or undershirts.

In contrast, an evolving supply process is where sources of supply are rapidly changing, the manufacturing process is in an early stage, and the underlying technology is quickly evolving. Examples include alternative energy sources, certain organic food items, or high-end computer technology. It is challenging to manage supply chains with either demand or supply uncertainty. It is especially challenging to manage supply chains with *both* uncertainties (see Figure 7.3).

Demand Uncertainty

	LOW *Functional Products*	HIGH *Innovative Products*
Supply Uncertainty LOW *Stable*	Efficiency Focused SC ***Least Challenging***	Responsive SC
HIGH *Evolving*	Risk-Hedging SC	Agile SC ***Most Challenging***

Figure 7.3 Demand and supply uncertainty

7.2.1a Efficiency-Focused Supply Chains

Efficiency-focused supply chains have both low demand and supply uncertainty and are easiest to manage. These supply chains typically don't have high profit margins. However, their operations are highly predictable and the gains in these supply chains come from efficiency and elimination of waste.

What to use here? The efficiency focus means using big data analytics to automate and standardize the order-processing cycle, to rely on electronic auctions, and to use simple visualization and dashboards.

7.2.1b Responsive Supply Chains

Responsive supply chains are used for innovative products that have a stable supply base. The primary challenge here is ensuring that you are able to quickly respond to customer demands. Mass-customization strategies such as *postponement*, where the supply chain is designed to delay product differentiation as late in the supply chain as possible, are effective here. A good example is the case of HP printers that are kept in generic form as long as possible. They are differentiated with country- and language-specific labels that are added at the last point in the system.

What to use here? Analytics applications can be very effective in designing postponement into the manufacturing and distribution process.

7.2.1c Risk-Hedging Supply Chains

Risk-hedging supply chains are those with high uncertainty on the supply side. These supply chains must do everything possible to minimize risks of supply disruptions and inventory shortages. These supply chains typically rely on higher inventory safety stocks and engage in a practice where resources are pooled and shared between different companies.

What to use here? Analytics applications for scenario planning are important here.

7.2.1d Agile Supply Chains

Agile supply chains are used in cases of both high demand and supply uncertainty. They are the most difficult to manage as they simultaneously must be responsive to an uncertain demand while using strategies to hedge risks to ensure there are no supply disruptions. These supply chains use mass customization strategies on the demand side, while carrying higher safety stock and engaging in resource pooling on the supply side.

What to use here? Here, scenario planning and predictive analytics for risk management and supply chain disruptions are important.

7.2.2 Single Versus Multiple Sourcing

The traditional view of many companies was that multiple sources of supply were best. This would increase cost competition and ensure supply security. This view has been challenged for quite some time and the notion of single sourcing has becoming the acceptable norm. Single sourcing focuses on building closer supplier relationships and cooperation between buyers and suppliers, and moves away from arm's-length relationships. It focuses on moving away from competitive bidding, using cooperative negotiation, and building long-term relationships.

Single sourcing has a number of benefits. Splitting the order among multiple suppliers can be costly as it doesn't permit consolidating purchase power. Splitting the order also has an impact on quality as there will be natural variations between different sources, even if minimal. Single sourcing also lowers freight costs and enables easier scheduling of deliveries.

In some cases, there might not be a choice but to go with a single source. This may occur, for example, when a supplier is the only source of a particular material. They may be the only one that has the needed process or they are the sole owner of the patent for the product. Having a single source of supply, however, is risky in practice. When a supplier is small, the buyer's business may utilize most of the supplier's capacity. This makes the supplier highly vulnerable if there is a discontinuity of purchase from the buyer. At the same time, the buyer might not want to be tied to a source that is so highly dependent upon them. Similarly, by relying on one supplier, the buyer is vulnerable if there is a disruption in the supplier's production process, such as a fire at a plant, a labor strike, or a work stoppage if the supplier's supplier has a problem.

The best strategy is to use a portfolio of a small number of multiple suppliers. Similar to a financial portfolio, these suppliers can be selected to create a balance of requirements. Some can be local for rapid deliveries while others may be global but less expensive. This strategy can help balance the risks of relying on a single supplier in case of supply chain disruptions. Another rule to consider is that one buyer should not make up more than 20 or 30 percent of the total supplier's business, otherwise making the supplier highly vulnerable.

Optimizing the bundle of suppliers that balances risks and costs is best done with analytics applications. Amazon, for example, does just that. It uses analytics for every supply chain lever—including sourcing.[13] The company uses analytics to determine the optimal sourcing strategy and manage all the logistics to get a product from manufacturer to customer. This includes using analytics to determine the right mix of joint replenishment, coordinated replenishment, and single sourcing. In fact, Amazon applies advanced optimization and supply chain management methodologies and techniques across its fulfillment, capacity expansion, inventory management, procurement, and logistics functions. These are then linked to all the other supply chain levers.

7.2.3 Domestic Versus Global Sourcing

Another challenge to consider is whether to use domestic or global sourcing, also called *offshoring*. Global sourcing rapidly grew as companies were attracted to cheaper labor costs in other parts of the world. It has been most prominent for sourcing products with easily defined standards, such as in the retail industry. It is also used heavily in the service industry, such as running call centers, processing claim forms, or in software development. Even in medicine, reading of diagnostic tests, such as X-rays, is often offshored.

However, with a rise in fuel prices, the labor savings are often negated, or even outweighed, by high transportation costs. Recent years have seen a trend of *nearshoring* or *backsourcing*.

A number of companies are turning toward domestic sources to reduce transportation costs, monitor quality more closely, and have closer buyer-supplier relationships. Also, with an emphasis on sustainability and "green," there has been a push toward local and regional sourcing.

Optimizing the mix of global versus local sourcing is most successful through big data analytics. IBM's supply chain picture was very different two decades ago.[14] It had a supply chain structure suited to supporting regional product sales across 150 countries, with different business units handling sourcing, logistics, and delivery of orders. "We had local procurement, local cash collection, local unique processes, and many units had their own [information] systems," Carroll recalls.[15] Then IBM recognized that a supply chain strategy focused on local or regional businesses was no longer viable. The company moved toward global delivery of software and services. In 1993, the company began the process of reorganizing its many supply chain organizations into a single global entity. The first step was to transform its procurement and order fulfillment functions, including establishing standards for those activities for all business units in every country. IBM established Global Sourcing Councils where procurement executives could exchange knowledge with their counterparts in other countries. Analytics enabled global standardization as well as the ability to solve problems and coordinate with other functions.

7.3 Outsourcing

Before deciding to outsource certain business functions, a company should ask itself: why outsource in the first place? Analytics is a new competency, and businesses should think carefully before outsourcing this driver of business intelligence. Companies should also set boundaries in terms of what they will and will not outsource.

7.3.1 Why Outsource?

Outsourcing has become a mega trend in many industries and is continuing to grow, as companies focus on their core competencies and shed tasks perceived as noncore. Companies may choose to outsource activities or tasks for many reasons, rather than performing them internally. These include lower costs, access to technical skills, and the ability to free themselves of doing noncore activities. Outsourcing enables companies to tap into specialized skills and capabilities that they currently do not have.

A good example has been the growth in outsourcing of transportation of goods to third-party logistics providers (3PLs), such as UPS and FedEx. Transportation and movement of goods requires investment of specialized resources, such as fleets of trucks, aircraft, and state-of-the-art information systems for product tracking. Outsourcing this activity to specialty companies is a much more cost-effective decision for most firms. Good outsourcing decisions can result in lowered costs and a competitive advantage, whereas poorly made decisions can increase costs, disrupt service, and even lead to business failure. Although financial aspects of outsourcing are important, outsourcing has increasingly taken on a broader strategic organizational focus.

It is for these same reasons that we are now seeing the rise of outsourcing in analytics services. We are seeing the rise of the third-party analytics providers (3PAs) as more companies recognize the symbiotic benefits that can arise from working in concert with an external provider. Technological capability is so highly specialized and growing exponentially that companies simply cannot keep up. All indications point toward the continuation of this trend in the foreseeable future.

7.3.2 Analytical Outsourcing

Analytics is not part of the traditional core capabilities of most companies. To utilize the technology and tap into its full potential, firms need to somehow acquire it. Many are working aggressively to hire analytical talent. However, for most companies that is not enough. Companies need deep analytical capability that requires high specialization. They need analytical professionals to analyze data and develop models. They also require employees who can do some analysis but can communicate with those on the "front lines." These companies need individuals with analytical skills who can also communicate. These individuals are hard to come by.

There is a general consensus as to the lack of skilled employees required to manage and utilize analytics. It is especially difficult to find analysts who understand a range of techniques, not just one. This is a problem, as integrating overlapping analytical technologies is increasingly necessary to tap into its full capability. The effective use of analytics depends not only upon the availability of data and analytical tools, but also upon the ability of companies to use them effectively. Without such talent, it is not possible.

Analytical outsourcing arrangements are necessary as companies do not have the skill set to acquire and keep up with this technology. Most companies lack the capabilities to do all the analytical work that is required. External providers often have access to data and software and are on the technology frontier. In addition, many providers of software and data are also now providing analytical assistance. Their expertise on key data sources and analytical techniques makes them uniquely qualified to help with analysis. Many external providers specialize in certain industry segments and can bridge the gap between industry and technical knowledge.

Examples of such outsourcing arrangements abound. Accenture, for example, provides consulting and outsourcing services to a variety of companies, including Best Buy. Accenture assists with analytical strategy and with particular analytical applications, and in Best Buy's case also manages major components of its IT function.[16] Software firm Teradata worked closely with Hudson's Bay Company to implement a new approach to reducing fraud in the merchandise returns process. Alliance Data works with retailers, such as Limited Brands and Pottery Barn, to establish and manage their loyalty programs. Mu Sigma provides analytical services to Walmart and other retailers. Based on these examples, it is clear that even leading analytics companies outsource at least some aspect of their analytics capability.

7.3.3 How Much to Outsource?

Outsourcing provides tremendous benefits—particularly enabling an organization to tap into highly specialized skills it currently does not have. However, outsourcing has inherent risks. Outsourcing to an outside vendor can create a range of risks and dependencies, not the least of which is the leaking of proprietary information. Companies need to give a great deal of thought to the extent of their outsourcing engagement, considering both immediate and long-term benefits. Outsourcing is not a simple "make-buy" decision. It needs to be a strategic decision carefully considering the skills the company will keep internally versus those it will access externally.

Two key dimensions that help define outsourcing are the *scope* and *criticality* of the outsourced task.[17] Scope is the degree of responsibility assigned to the supplier. Criticality, on the other hand, is the importance of the outsourced activities or tasks to the organization. The greater the scope of the outsourced task, the larger the relinquishing of control by the buyer. Similarly, the greater the criticality of the outsourced task, the greater the consequences of poor performance to the buying firm and greater the requirement for supplier management. It is important to understand these dimensions when evaluating the outsourcing decision.

Outsourcing can involve just one aspect of analytics. For example, the outsourcing engagement can be small in scope—such as purchasing data or conducting analysis on one data set. It may mean supplementing current staff with either onshore or offshore analytical consultants. On the other hand, it may mean outsourcing most of the company's analytics capabilities. Companies should be especially careful when outsourcing activities with large scope and criticality, and should especially monitor such outsourcing engagements to ensure a successful outcome.

The outsourcing trend will continue to increase as technology continues to evolve requiring ever-specialized skills. As described in Chapter 2, "Transforming Supply Chains," a well-formulated analytics outsourcing strategy is a must. This strategy must clearly delineate which analytical capabilities the company wants to build for itself and which it will outsource from partners. It must also specify the long-term plan of building capabilities over time, in addition to meeting short-term goals (see Figure 7.4).

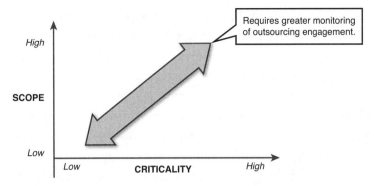

Figure 7.4 Dimensions of outsourcing

7.4 Risk Management

Big data analytics has application in risk management. It can be used to analyze the risk of suppliers, monitor security risks, and manage supply chain disruptions.

7.4.1 Analyzing Supplier Risk

Success of supply chain operations requires a reliable and robust supply network, from external suppliers to contract manufacturers.[18] Given the high dependence on suppliers, companies need to evaluate supplier performance and risk on a routine basis. Supplier risk analytics is still underdeveloped compared with other supply chain areas. In most organizations, supplier risk analytics still involves simple metrics and reports. However, there are numerous examples of where big data analytics can improve supplier risk assessment.

One example of the use of big data analytics is the creation of a *supplier resiliency score*, which can use many variables to identify high-risk areas.[19] The variables can be determined based on expertise and understanding of problem areas. An example is high risk of weather events near a supplier's manufacturing location or the availability of alternative production sites. If the variables or the overall resiliency scores suggest a problem, companies can then pursue alternative sourcing or work with existing suppliers to identify contingency plans or alternative locations. Cisco is a case in point of a company effectively using a supplier resiliency scorecard.[20] The company faces significant risks as most of its manufacturing activities are outsourced. As a result, the company relies on a resiliency scorecard that includes four categories—manufacturing resiliency, supplier resiliency, component resiliency, and test equipment resiliency. The scorecard identifies areas with the highest risk and helps Cisco take action to remedy the potential problems.

A second area for analytics is the evaluation of supplier performance based on how they handled recent events and risks. Examples range from how the companies handled an economic downturn to a supply chain disruption. Analytical tools can incorporate public third-party data and help companies assess this risk.

A third area of supplier risk assessment relates to human trafficking and labor laws. The United States Department of Labor reports that more than 100 common household items are produced using either forced labor, child labor, or both.[21] Such items include tea, coffee, cocoa, garments, textiles, precious metals, and others. Although the topic of forced labor and child labor conditions has been studied within social science disciplines, there is a severe shortage of best practices for analytic modeling and decision analysis to address such conditions. The groundbreaking 2010 California Transparency in Supply Chains Act legislation requires retailers and manufacturers to evaluate and address risks associated with slavery and human trafficking within global supply chains. In addition to having to meet legal requirements, the existence of forced labor conditions may have huge impacts on organizational strategy, policies, and brand image, and may introduce a variety of other risks.

Big data analytics plays a large role for companies. It can help identify high-risk areas for forced labor within supply chains. It can also help develop decision models to prioritize resources for risk identification and reduction. It can also help develop capabilities to improve detection of forced labor conditions.

Intel is a good example of a company using big data analytics for risk mitigation.[22] The company uses big data analytics well beyond tactical supply chain decisions. It helps the company look at broad issues such as supply chain compliance, social responsibility, and sustainability. By monitoring social media, for example, the company can learn about a supply management problem in a remote location or a supply base. Through its own data capture, the company can conduct analysis to proactively begin to investigate and uncover problems. It can connect social media information to other areas of information to determine whether the problem is with one of its suppliers—or if it's a problem with the supplier's supplier. Intel models and simulates various supply chain scenarios and lets the data offer insights as to how to make improvements on the supply side.

7.4.2 Security

Data-driven supply chains have become more sophisticated in utilizing information technology and software to manage and coordinate end-to-end operations. This technology, however, has made supply chains more vulnerable to delays and disruptions arising from various internal and external factors. Security is the most significant risk facing supply chain organizations.

Security—whether product or data security—has never been more important. Today, this goes well beyond the traditional concerns of product theft and pilferage, and is now coupled with more recent issues related to terrorism or large-scale piracy. It has been estimated that security breaches in the United States cost tens of billions of dollars each year.[23] The Department of Homeland Security has implemented the Container Security Initiative (CSI) to combat terrorism's potential impact on cargo arriving in the United States. And the problem is significant. It is estimated that more than 12 million containers arrive at U.S. ports every year.[24]

For companies such as FedEx, security is essential to the business. The company handles nine million shipments a day and manages all the accompanying data. The company recently embarked on a mission to get the full potential out of the large stores of data it had. FedEx decided to apply that data to physical items. It created a next-generation, first-of-its-kind information service called SenseAware that combines a GPS sensor device and a Web-based collaboration platform.[25] Originally used by the health-care and life sciences industries as a means to track high-value and/or extremely time-sensitive shipments (and now available to all industries), SenseAware attaches digital information to packages, providing information such as precise temperature readings, information about a shipment's exact location, notification when a shipment is opened or if the contents have been exposed to light, and real-time alerts and analytics between trusted parties regarding the above vital signs of a shipment. Because the device is equipped with a radio that constantly broadcasts information back to FedEx, an enormous amount of data is generated—information that must be acted on in real time. This has provided top-level security and efficiency and generates data for large-scale analysis.

7.4.3 Supply Chain Disruptions

Coca-Cola has always been more focused on its economic bottom line than on global warming. Then in 2004, the company lost a lucrative operating license in India because of a serious water shortage.[26] Global droughts have continued and repeatedly dried up the water needed to produce the company's soda. After a decade of struggle, the company has embraced the idea of climate change as an economically disruptive force to its supply chain.

"Increased droughts, more unpredictable variability, 100-year floods every two years," said Jeffrey Seabright, Coke's vice president for environment and water resources, listing the problems that he said were also disrupting the company's supply of sugar cane and sugar beets, as well as citrus for its fruit juices. "When we look at our most essential ingredients, we see those events as threats."[27]

A report by consulting firm, WisdomNet, Inc., identifies a number of factors that are contributing to the vulnerability of organizations and their supply chains.[28] These include globalization resulting in longer supply lines, an increasing trend toward outsourcing, which makes "others" responsible for what goes on in the supply chain, and lean operations that have removed inventory from the system, which make firms more efficient, but more vulnerable when unforeseen events occur. Just one of the many forces are natural disasters. Coke reflects a growing view among American business leaders and mainstream economists who see global warming as one force that contributes to lower gross domestic products, higher food and commodity costs, broken supply chains, and increased financial risk. These factors will continue into the foreseeable future.

The dispersion of sources of supply coupled with a lean pipeline is especially problematic. Disruptions can have a devastating effect as there are no buffer inventories and the lead time is long. Also, given the interconnectedness of global supply chain, disruptions in one part of the globe can easily propagate through the network. This can have deleterious effects on the organization, from drops in stock valuations to shutting down.

Consider the Swedish telecom company, Ericsson. The company lost the equivalent of USD $400 million in sales when a computer chip supplier in the United States had a fire that closed its facility.[29] There was no backup supplier available in the short term, and the company could not meet demand for its mobile telephones for most of the year, leading to a loss of USD $1.86 billion for the year.

The key message is clear. There are many risk factors that can negatively impact the ability to serve customers. Companies understand that the biggest challenge they are facing is to protect the enterprise, the clients, and their shareholders from the unknown. These leading-edge companies rely on predictive analytics tools to help them foresee problems and take preemptive actions to prevent supply chain interruptions anywhere in the world. The global, integrated supply chain organization will ensure that those actions are carried out quickly, efficiently, and consistently, no matter where or when they're needed.

Part III

The Framework

8 —————————

The Roadmap

In 2003, Rocky Mountain Steel Mills—a leader in steel fabrication and construction—was forced to shut down pipe milling operations due to price pressures.[1] Then in 2005, the cost of oil had increased. The pool of potential customers and oil drillers grew rapidly and the company had to reconsider its strategy. Should it reopen the pipe mill? If so, when? Should it start production right away? Should it start taking orders now or wait until after reopening?

Demand, pricing, production constraints, and industry capacity are all factors that impact such strategic decisions. Standard costing and volume analyses are typically too simple given the complexity of these types of decisions. Complicating matters for Rocky Mountain was that managers and potential customers all vocalized strong opinions on the matter. Everyone had an eye on booming oil prices and urged production. Rocky Mountain's answer, however, was to rely on big data analytics to inform the decision.

The company chose a software application that could be used to conduct what-if analyses, such as forecasting profitability given a variety of factors. The company was able to use big data to perfectly time the reopening of production while avoiding market risks and maximizing profit. The data-driven decision was to delay reopening the plant, and then delay order intake even after production had commenced. This enabled the company to take advantage of price increases that were successfully predicted by the analytical software. As a result, the company enjoyed profitability and an increase in stock price.

8.1 Lessons

It is easy to assume that companies leading in big data analytics, such as Rocky Mountain, are simple number crunchers. And yes, it is true that they often apply technology with brute force to solve business problems. However, making that assumption leads you in the wrong direction. These leading companies do much more to be able to leverage big data analytics. They have the right *focus*, their efforts are *coordinated*, and they build the right *culture* that enables them to make optimal use of the data they are analyzing.[2] Leaders must remember that it is people and strategy, as much as information technology, that are needed to achieve this level of competitiveness.

8.1.1 Focus

Analytics competitors depend on data-driven decisions. However, they must choose *where* to direct resource-intensive efforts. It is not possible to effectively focus on everything. In fact, big data implementation that tackles too many areas can easily become diffused. It can lead companies to lose sight of the business purpose behind the effort. Highly focused efforts are an important lesson.

Leading companies focus their big data efforts by strategically selecting areas or initiatives that support their overarching strategy. UPS, for example, had initially focused their analytics efforts on improving logistics operations.[3] The company has since expanded its analytics efforts to provide superior customer service. Similarly, Harrah's Hotel and Casino has focused its big data analytics efforts on improving the customer experience.[4] This meant engaging in targeted efforts to increasing customer service and areas directly related to the customer experience, such as pricing and promotions.

8.1.2 Coordinate

Leaders in big data analytics know that they must coordinate their effort internally, and with customers and suppliers. Recall that

Walmart requires its suppliers to use its Retail Link system to monitor product movement by store, to plan promotions and layouts within stores, and to reduce stock-outs. Similarly E.&J. Gallo provides distributors with data and analysis on retailers' costs and pricing.[5] This enables the distributors to calculate the profitability for each of Gallo's 95 wines. The distributors, in turn, use that information to help retailers optimize their assortment on the shelf and persuade them to add shelf space for Gallo products.

Then there is Procter & Gamble (P&G), which offers data and analysis to both its retail customers and suppliers. This is a program called Joint Value Creation, which has helped improve responsiveness and reduce costs.[6] Similarly, hospital supplier Owens & Minor offers customers and suppliers the ability to access and analyze purchase data, such as tracking ordering patterns in search of consolidation opportunities.[7] Using the system, for example, a hospital may be able to evaluate cost savings that could be achieved by consolidating purchases across multiple locations. They can also assess tradeoffs between order frequency and carrying costs.

The lesson here is this: To reap the greatest benefits from big data analytics, companies must work along the entire supply chain. They must link their data and applications with customers and suppliers. They must also measure performance of their supply chain partners as well as their own. A supply chain is a system, as discussed in Chapter 2, "Transforming Supply Chains," and must be optimized as such. Big data analytics can make this happen.

8.1.3 Culture

Companies that succeed with big data analytics *do not* do so because they have more or better data. Rather, companies succeed because they have leadership that creates the right culture, is able to set and articulate clear goals, and gets everyone "on board." They are able to effectively lead their organization through transformational change; they create a culture where decisions are based on data and facts. Questions such as "what do we know?" are used to replace the old "what do we think?"[8]

Certainly availability of data and analytical tools are important. The effective use of big data analytics, however, depends upon the ability of executives, managers, and employees to use analytics effectively. The power of big data does not take away the need for leadership and human insight. Leaders still need to identify strategy and opportunity, understand markets, articulate a vision, and think creatively. Successful companies are the ones whose leaders do all these things while creating an organization that is data driven.

8.2 Doing It Right

Big data offers a transformative potential for supply chain management, along all supply chain levers, from Source to Sell. But how do companies leverage this for a competitive advantage?

To achieve a competitive advantage, these efforts need to be focused and coordinated, following a three-step roadmap: segment, align, and measure (SAM). Following the roadmap enables the pieces to be aligned, rather than remain pockets of excellence.

The process begins with using big data analytics for supply chain *segmentation* that enables definition of competitive priorities for each segment. The process then *aligns* organizational functions to support competitive priorities and engage in analytics efforts to support this vision—rather than random exploratory efforts. Alignment serves to integrate horizontally throughout the entire organization and supply chain by sharing "intelligence" across functions through joint decision making, such as Sales & Operations Planning (S&OP). Lastly, these initiatives are measured with *targeted and measurable key performance indicators (KPIs)*. Otherwise, these efforts will remain isolated pockets that do not have large scale and lasting impact (see Figure 8.1).

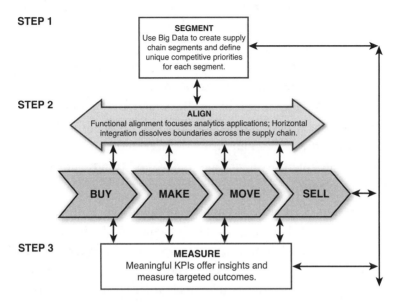

Figure 8.1 The roadmap—SAM

Figure 8.1 illustrates that big data analytics implementation should follow a systematic framework across the supply chain called SAM. It begins by using analytics to create supply chains that target specific market segments with clearly defined competitive priorities. Next, strategic alignment across functions and integration of intelligence across the organization and the supply chain leads to optimal use. Last, well-defined metrics enable assessment of performance. A feedback loop enables continuous improvement.

This framework provides the roadmap and links the three steps.

8.2.1 Segment

Segment: Create optimal segmentation with clear attributes.

The first stage of the roadmap focuses analytical implementation efforts on the most important supply chain activities. It helps companies *avoid the trap of searching for a needle in a haystack* or reading into meaningless correlations.

Segmenting and analyzing customers—by analyzing attributes such as demographics, customer purchase metrics, and shopping attitudes and behavior—have been used for decades. Segmentation enables dividing the target market into subsets of customers who have common needs and priorities. These segments are served through different channels, have different products, and have different supply chains. However, big data enables data scalability. Granular data can be aggregated in an infinite array of possibilities. This makes it possible to create microsegments and understand trends and outliers within each segment.

A critical part of creating segments is defining competitive priorities in each segment. Defining unique segments and their characteristics leads to a clear identification of competitive priorities for each segment. Competitive priorities identify how a company competes in each segment. They include variables such as customer service, cost, quality, time, flexibility, and innovation. Each of these then translates into different operational requirements.[9] This results in different supply chain structures, suppliers, transportation, operational strategies, and thresholds of performance for each segment. Consider that supply chain segments that focus on cost will have very different supplier metrics than those that compete on innovation, quality, or customer service. Each segment focuses on different objectives. The goal is to identify the best supply chain processes and policies to serve each customer at a given point in time while supporting the driving business strategy.

8.2.2 Align

Align: Align efforts to support segment attributes.

Alignment throughout the organization and supply chain avoids fragmented efforts.

Strategic functional alignment should drive analytics applications rather than engaging in fragmented efforts that do not support competitive priorities. Otherwise, investments into this emerging IT competency will produce fragmented efforts. Without alignment, all the data mining in the world will not yield a competitive advantage. *Alignment avoids fragmented efforts.*

Aligning means integrating processes across the supply chain. Leading companies are now using predictive analytics to dissolve the boundaries between customer relationship management, on the Sell side of the supply chain, and operations, procurement, and logistics in order to sync up supply and demand. Big data can be a huge source of aid in this process, enabling *demand sensing* and driving other supply chain decisions. Consider, for example, that Ford uses big data analytics to collaborate throughout its supply chain. FordDirect provides an interface to share information between the customer, dealer, and manufacturer in real time to customize vehicles, manage inventory, and obtain financing.[10] The sharing of this information enables integration and coordination across the entire supply chain.

S&OP implementation is especially useful to integrate decision making across the organization, as evidenced by leading companies such as P&G, Merck, Hershey's, and others. S&OP is a business management process intended to match supply and demand through functional collaboration; it relies on data and analytics to develop cross-functional supply chain policies including risk management. Integration means synchronizing production schedules with customer demands based on real-time market dynamics. Without integrating predictive data analytics into S&OP, manufacturers risk overstocks and stock-outs.

8.2.3 Measure

Measure: Develop strategically aligned KPIs to measure segment attributes.

As Peter Drucker said, "If you can't measure it, you can't manage it." An organization needs to look at the right metrics for the phenomena it needs to optimize. This can be accomplished using strategically aligned KPIs agreed upon by all process members and a feedback mechanism that enables continuous improvement. These metrics should also measure alignment, integration, and cross-enterprise cooperation. However, companies should use analytics to look for new meaningful analytics metrics, which are driven by strategy, core competencies, and understanding of the value proposition of the business. Big data analytics enables the development of new metrics that offer greater insight. Just consider the movie *Moneyball* where traditional baseball metrics to evaluate players, such as "batting average," were changed to new and more meaningful metrics, such as "on-base percentage."

8.3 How It Works

The roadmap links the key elements of implementing big data analytics across the supply chain, tying the strategic concepts with operational levers.

The first step is to use big data to create better supply chain segments. The idea of segmenting and analyzing customers through combinations of attributes—demographics, customer purchase metrics, and shopping attitudes and behavior—is firmly established. However, big data analytics takes this to an entirely new level. Big data enables microsegmentation for real-time analysis. The goal is to develop segments that optimize customer needs and supply chain requirements to serve each segment. Consider the apparel retailer American Eagle Outfitters. The company used big data analytics to create clusters of its more than 750 stores based on the types of assortments to which shoppers were most responsive.[11] The company was able to identify that customers in Western Florida bought merchandise similar to those in parts of Texas and California. This segmentation enabled the company to localize store assortments by segment and location cluster, better control pricing in each segment, and move resources into most promising segments in real time.[12]

Big data analytics can then be used to define clear competitive priorities for each segment. Competitive priorities all translate into different operational requirements—for example, focusing on cost for low-margin segments versus customer service in high-margin segments. This results in different supply chain structures, suppliers, transportation, operational strategies, and thresholds of performance. Analytics algorithms can then optimize decision processes based on the competitive priorities in that segment—such as optimizing customer service while keeping a threshold on costs. This provides focus. Algorithms can then automatically fine-tune inventories and pricing in response to real-time, in-store and online sales—targeting the set competitive priorities. Manufacturing companies can adjust production lines automatically to optimize efficiency and reduce waste.

Next align the organization and all business functions to support the defined competitive priorities. Implementation of big data and analytics should support competitive priorities in each segment, rather than being randomly implemented. This helps focus implementation rather than using a "finding a needle in a haystack" approach. You also need to integrate across the enterprise to match supply and demand. A process such as S&OP is an excellent place to introduce big data analytics as the process itself is data dependent and crosses functional boundaries.

Lastly, you need to measure performance and outcomes using strategically aligned metrics or KPIs. It is important here to use metrics for *continuous improvement*. A feedback loop should exist between metrics—monitored on a continuous basis—and the defined segments and their competitive priorities. The metrics should be used to refine the segmentation process and realign the competitive priorities. Through Total Quality Management (TQM) and *kaizen*—a Japanese term for continuous improvement—we have learned that the best and sustained improvements occur from gradual improvements.[13] This should be an ongoing process. Big data algorithms can significantly assist with this, such as automatically tracking these metrics and creating alerts when deviations occur.

8.4 Breaking Down Segmentation

Segmentation—step 1 in the roadmap—enables companies to know their core markets and competitive priorities in each segment. Big data enables the creation of customer microsegments based on aggregation and disaggregation, with detailed insights into the unique needs of each segment. *The result is optimal segmentation with clear attributes.*

8.4.1 End of Standardization

Big data analytics has created an end to the era of standardization. Segmenting involves tailoring every possible aspect of the product offering to highly specific markets. In retail, for example, analytics applications are used to conduct what is called cluster analysis. Cluster analysis is a statistical technique that has been used for decades. Big data, however, takes this to a much greater level. This has enabled creating groupings or clusters of similar stores—called *clustering*—based on a wide variety of characteristics. The analytics applications then identify the best merchandising and sales strategies for each particular cluster.

Analytics can even be used to tailor offerings to individual stores, rather than clusters of stores. The applications can consider a variety of attributes unique to that store, including population demographics and shopping behaviors of local customers. This is called *localization.* The application can then specify the best attributes for that unique store at that location, including product assortments, pricing, store formats and layouts, promotions, and even staffing levels.

This is a huge trend in retail. Simply put, companies understand that the demand for their goods—and the impact of promotions and store layouts—varies by geography and customer demographic. In the past, standardization was an economical approach from an operations standpoint. It allowed companies to achieve economies of scale. However, big data analytics has changed that. Companies can now use big data analytics to localize as much as possible. They can now do it in an economical way. The result is increased sales and profits.

8.4.2 Focused Competition

An important aspect of segmentation is defining *competitive priorities* for each segment—the primary way of competing in that segment. Understanding segments leads to clear identification of competitive priorities for each segment. Competitive priorities—customer service, cost, quality, time, flexibility, and innovation—all translate into different operational requirements.[14] Recall that different competitive priorities need to be supported by different supply chain structures, supplier selection, transportation modes, and thresholds of performance for each segment.[15] Segmentation lets companies tailor their supply chains to each customer and product in their portfolio based on how they compete in that segment. This stage of the roadmap focuses analytical implementation efforts on the most important supply chain activities. It helps companies *avoid the trap of searching for a needle in a haystack* or reading into meaningless correlations.

Just consider the following. It would not make sense for a company competing on low cost to have its first analytical implementation focus on improving customer loyalty and customer relationship management. The reason is pretty obvious—efforts do not match the competitive priority. Also, low-cost providers typically have small margins. As such, they would likely not be able to afford the information and analytical infrastructure—people and technology—needed to mount a major analytical initiative. For such firms—focused on low cost—applying analytics efforts toward supply chain and pricing would be more logical.

Similarly, it would not be reasonable for a high-end retailer with few select store locations to target their first analytical effort on supply chain and site selection analytics. Given the competitive priorities of these types of firms, it would be more logical to focus on customer-oriented capabilities. It is important to remember to let the competitive priorities in each segment dictate the analytical focus.

The broad range of possible analytical tools that companies can adopt makes it essential for executives to understand which analytical activities match their strategies and organizational capabilities. Companies cannot pursue the full spectrum of possibilities with equal rigor. Recall that big data efforts at leading companies are conducted in a focused manner. Therefore, to be successful, companies need to use analytics in a targeted fashion to support their strategies, business models, or organizational capabilities.

Companies such as Brooks Brothers, Nordstrom, and Nieman Marcus have a primary focus on developing and maintaining long-term customer relationships. As a result, these firms have a computerized system in place to capture customer interactions and manage customer relationships, called *clientelling*.[16] A good strategy for these companies may be to add analytics capabilities into the clientelling system, enhancing their primary focus.

Electronics retailer Best Buy has used big data to develop clear market segments and locations, understanding its competitive priorities.[17] The company uses this information to customize store layouts, product assortments, and service approaches for each location. The segment-based localization approach also involves the extensive training given to store managers and personnel on how best to serve each segment. Best Buy is now beginning to localize store formats based on area demographics and customer behavior data.

A few decades ago, Walmart realized that supply chain analytics were the foundation to their success. These tools were critical for maintaining high product availability while keeping costs low—a hard combination to attain. One area it has used is site selection analytics. The company has more than 2,500 Walmart and almost 600 Sam's Club stores.[18] As a result, optimizing site selection is critical. Now, in addition to maintaining these capabilities, the company is using its analytical capability toward customer behavior analysis and the shaping of customer demand.

Using big data analytics to segment markets and customers has not been restricted to retail. Insurance companies and credit card issuers that rely on risk judgments have been using big data to segment customers for many years. In fact, the entire value proposition of Progressive Insurance is that it uses data analytics to create customer segments and optimize insurance offerings and pricing.[19] Further, market segmentation can include a social network, rather than typical demographic factors. eBay, for example, uses social network theory to identify social clusters and uses that information to target advertisements.[20] First, the company uses online tracking technologies to identify customers shopping on its site. Then—with the help of the start-up 33Across—it analyzes data from social networking sites to map out the connections between the customer it identified and other Web surfers. By using cookies and other tracking devices, it can tag these customers and use that information to target ads at the appropriate time. Sprint used this approach to launch its Palm Pre smartphone and quadrupled related online sales.[21]

8.5 Strategic Alignment

Analytical applications need to be aligned with strategic drivers and competitive priorities.[22] Alignment gives us a plan, direction, and focus. Without alignment, investments in big data analytics produce fragmented efforts. Today's analytics application can optimize anything. The question is: Does it matter?

UK-based retailer Tesco is a good example of a company that aligns its big data efforts along the entire value chain to realize its fullest potential. The company creates analysis of big data from Club-Card, its successful loyalty program.[23] The loyalty program helps the company gather purchase data. Tesco then analyzes the data to inform a variety of decisions, conducting microsegmentation of its customers and optimizing product mix, pricing, and promotions. Based on data analysis, the company then aligns organizational efforts toward each segment, such as tailoring store formats to particular locations, ranging from convenience stores to online stores. As a result, Tesco can determine exactly what types of customers buy from each store format, and what products they buy, and ensure availability of these products when and where customers want them.

8.5.1 Matching Supply and Demand

One of the biggest issues for supply chain management is matching of supply and demand. Big data analytics is extremely useful to achieve this balance. Analytics enables demand sensing on the demand side and enables driving other supply chain decisions on the supply side. An excellent example of how this works is the toy chain Toys R Us.[24] The company uses a time-phased approach to replenishment that gives them—and their suppliers—visibility into the continuous stream of future product needs. The company continuously provides order projections to its suppliers, linking demand projects with supply. As a result, Toys R Us reduces its own inventory risk, as planners at all levels are able to be proactive in resolving potential problems before they arise. By focusing on demand and replenishment at the store and at the SKU level, the company puts the emphasis on the customer rather than simply pushing products into its supply chain. Synchronization of demand and supply between stores and distribution centers has resulted in lower inventory levels while at the same time reducing stock-outs.

Another example of matching demand and supply through big data analytics can be seen in the auto industry. Data from dealerships can be used to modify production schedules and avoid overstock. It also enables matching buyers with available vehicles that fit customer-specific preferences. In the past, manufacturers had decided on the vehicle make, model, and specific comfort features ahead of production. Products that were fully finished were then shipped to dealerships. This occurred well before the customer had even expressed preferences—creating large risks for manufacturers.

Big data analytics has changed that. Big data analytics enables optimizing supply-and-demand forecasts and determining correlations between options, makes and models, and customer profiles. These insights can be used to match dealership orders with production schedules. They can allow customers to browse vehicle availability, design products to their preference, and evaluate the price. In automobile manufacturing, this can lower production costs, reduce the risk of surplus inventory, and enhance customer satisfaction.

8.5.2 Cross-Enterprise Integration

Successful supply chain management depends on cross-enterprise integration. Leading companies are now using predictive analytics to dissolve the boundaries between customer relationship management and supply chain management. Internally, the Sell side of the supply chain is now tied to the other levers—Make, Move, and Buy—in order to sync up supply and demand.

This type of integration is what the department store chain J.C. Penney initiated in 2001. The company began a major transformation of forecasting and planning, agreeing to employ common demand forecasts for the entire merchandising process.[25] Using big data analytics, the company fully integrated financials, sales, inventory plans, and in-season forecasts at the company and store level. Forecasts were also directly tied to assortments, allocations, and pricing optimization systems. The executives at Penney's credit the new integrated approach with an increase in gross margin, improvements in inventory turns of 10 percent, store sales growth for four consecutive years, and double-digit increases in operating profit.[26]

Another example of successful use of analytics for integration is seen at Deere & Company. The company committed to the use of big data analytics on the issue of product configuration.[27] Through analytics, the company identified the product configurations that were the least desirable to make and sell. This effort integrated both demand and supply. Analytics was used to identify less-profitable configurations, which were eliminated, resulting in reduced inventory and reduced complexity. The team of analysts then used analytics to determine the "ideal" product configurations for two existing product lines, creating reductions in configuration options while simultaneously maintaining customer satisfaction. The two product lines resulted in a 15 percent increase in profit.

8.5.3 Integrated Forecasting

Forecasts drive entire supply chains. As a result, forecasting is of critical importance to companies across all functions. Sales and demand forecasts drive marketing and operational and finance processes, including replenishment, promotions, real estate, budgeting, and even human resource decisions. Benefits of better forecasting include effective and efficient allocation of resources, reduced inventory stock-outs and excess goods, faster and more accurate management decision making, and enhanced coordination between functional groups, headquarters and stores, and with external suppliers.

Forecasting of sales and demand can be very difficult when there are thousands of different products and variants. This is particularly true for retail, which is characterized by constantly changing trends and consumer preferences, seasonal impacts on demand, thousands of different stores, alternative channels, and promotional influences on demand. The big data generated is rich with insights that have the potential to offer a more accurate view of demand.

In the past, many companies relied on manual forecasts. Having multiple manual forecasts not only takes more management time, but forecast differences make it difficult to coordinate operational and financial decisions across the retail enterprise. In many cases in the past, different groups and functions created their own forecasts to inform ordering, staffing, merchandising, and budgeting decisions. Stores and regions created bottom-up forecasts; corporate created top-down forecasts. Of course, the forecasts often differed and led to difficulties in integration and performance.

Today, companies are increasingly implementing integrated analytical forecasting techniques. A number of companies are both centralizing the forecasting function to create integrated forecasts and also moving to more automated and statistical forecasting approaches—rather than straightforward extrapolation—to rapidly generate forecasts with ranges of outcomes and probabilities. In addition, companies are looking to synchronize order forecasts across the supply chain—from demand signal through distribution centers and vendors—over an extended period of time. This delivers a time-phased accurate inventory order forecast based on consumer demand across the organization and the entire supply chain.

Contemporary forecasting analytical tools can generate millions of forecasts across organizational functions and be vertically disaggregated down to the store-item-day level. These forecasting tools can place confidence intervals around forecast results based on managerially set levels. They can also generate forecasts for special events such as holidays, can take weather and marketing promotions into account, and can generate a time-phased order forecast.

As more sophisticated automated statistical forecasting technologies are developed, companies will increasingly rely on integrated automated statistical forecasts to drive their decisions. Big data analytics will enable unprecedented levels of insights. However, implementing integrated forecasting does not only require technological initiative. It also requires a change in business processes and organizational roles. Most organizations have to reassign forecasting to a centralized organization to ensure that integrated forecasts are produced and employed throughout the organization. This also ensures the use of proper technical skills in generating these forecasts.

There is no question that these measures will substantially improve forecast capability for organizations and provide unprecedented visibility. However, there may also be political issues around centralized forecasting that senior management may need to address. Forecasts drive business decisions and require buy-in from those using them. Further, organizations need to ensure that these improved forecasts are linked to the decisions they drive—merchandising, supply chain, financial, human resource, and other processes. Without tying forecasting to actual decision making, forecast improvements will be yet another example of *islands of excellence*.

8.6 The Importance of Measuring

We have all heard some version of the standard performance measurement slogans: "What gets measured gets done," "If you don't measure results, you can't tell success from failure," "If you can't recognize success, you can't learn from it; if you can't recognize failure, you can't correct it," "If you can't measure it, you can neither manage it nor improve it." These slogans, or clichés, were developed because they underscore a reality of managing a business. They underscore the importance of measuring performance.

8.6.1 Metrics

A *metric* or *KPI* is any type of measurement used to measure some quantifiable component of a company's performance. Metrics are used to drive improvements and help businesses focus the organization—both people and resources—on what is important. They help determine if you are doing things correctly. Metrics indicate the priorities of the company and provide a window on performance, goals, and ambition. Ultimately, metrics help the organization answer the following questions:

- Where has the company been?
- Where is the company heading?
- Is something going wrong?
- Is the organization reaching its target?

To derive the most benefit from metrics, it is important to keep them simple. A common mistake with companies is to employ too many metrics and those that are simply too complex to understand. This is certainly true in the era of big data analytics when countless reports can be generated. The reports generated might look impressive, but it's possible that no one knows clearly where performance stands. Executives, managers, and all employees need to understand the metric, how they can influence it, and what is expected of them. For example, it is clearer to state that a metric's target is to reduce complaints down to two per month rather than stating that a metric's target is to have a 50 percent reduction per month. The latter is simply not specific. This communication element is a detail often overlooked, but it is important that employees have a good sense of what success might look like.

Good metrics do the following for an organization:[28]

- **Drive strategy**—Good metrics help measure the direction of the organization. They help the organization see if its strategy is working. Without metrics, the organization doesn't know if the strategy is effective or not.

- **Provide focus**—Good metrics provide focus for an organization, department, or employee. Metrics help focus employees' attention on what matters most to success.

- **Help make decisions**—Metrics should be explicitly defined in terms of unit of measure, who or what is being measured, collection frequency, data quality, expected values and targets, and thresholds.

- **Drive performance**—Good metrics are valid, to ensure measurement of the right things. They are also verifiable, to ensure data collection accuracy.

- **Are evolving**—These metrics change and evolve with the organization. Good metrics allow measurement of accomplishments, not just of the work that is performed.

- **Are clearly communicated**—Good metrics produce good internal and external public relations. They provide a common language for communication.

A primary purpose of using metrics and taking measurements is to assess performance levels and to analyze what is happening and where. The most beneficial aspect of performance measurement, however, is pinpointing problem areas and focusing attention on actions that will have the best impact on overall business performance.

Without good metrics or KPIs, it is easy to fall into the common big data hurdle of *measurement minutiae*. This is where you try to measure everything, not knowing what to focus on. This is the trap of having people busy with all kinds of activities, but achieving few measurable results. Effective performance measurement is the compass that guides management in a direction that will produce meaningful results at the process level, results that will tie directly to the company's goals.

Fact: It is very difficult to improve what you don't properly measure. And many companies focus on the wrong performance issues. The pressure to focus energy on activities that really matter must come from the highest levels of the manufacturing enterprise. Leadership can talk about the need for making improvements, but unless the right performance factors are measured and rewarded, nothing will change. Today's world-class companies are continually tracking process performance factors that ultimately impact business success, such as order-to-delivery cycle time, throughput, inventory levels, operating expense, and customer satisfaction.

Inappropriate measures often lead managers to respond to situations incorrectly and continue to reinforce undesirable behavior.

For example, when a company incentivizes marketing based on sales, but operations management focuses on keeping inventories low, there will be a mismatch between incentives—even in the era of big data analytics and integrated analytical forecasting. Customer service may suffer or inventory levels may be too high. Similarly, getting the lowest possible price is important, but ensuring an uninterrupted supply of needed materials to maintain the production schedule and meet customer deadlines is more important. Just think about the real cost of material shortages. The best purchased material value is a result of price, quality, and on-time delivery—*multiple metrics.*

Keeping an entire organization focused on the right targets and moving in the right direction is not an easy task. Often, what managers *think* their superiors consider important—through the formal or informal measurement system—is what is going to get done. People respond to metrics, so deciding what metrics will be used is an important task. For example, if top management pushes for the importance of lower inventories, the organization may focus on metrics that measure inventory levels without considering the impact on customer service. Remember that if the performance measurement system does not focus on a clear direction, the measurement system itself will enforce the wrong actions.

Conflicting performance measures will always result in the organization going in differing directions. Without uniform expectations, it is virtually impossible to keep an organization marching toward the same goals. This, by itself, makes reevaluating how a company measures business performance a top priority. Leadership must strive to direct all levels of its organization to focus on the right priorities. World-class companies have learned that effective performance requires linking the business strategy with day-to-day actions. However, it is measurement that will drive improvement and continue to encourage the right response and behavior.

Financial results are the ultimate measures, but are not drivers of business success. In fact, financial measures are typically a reflection of other performance variables. It is these performance variables that need to be measured directly for them to be improved. These other variables will ultimately impact financial performance.

8.6.2 Big Data Metrics

Big data metrics are primarily the same KPIs currently being used. The difference is that these metrics need to be brought out of their channel and functional silos and synthesized across the entire organization and supply chain. Also, these metrics must be strategically aligned.

Analytics applications offer tremendous opportunity for interpretation of metrics and enabling continuous improvement. Too many metrics can lead to information overload—where decision makers simply can't process all the information. The key is to present only what is relevant. This may be an executive wanting summary data across the company's product lines or a manager wanting more detail, but only for the areas that he or she oversees. IT professionals are challenged with not only providing the infrastructure, but also helping provide meaning to the big data. Data visualization techniques and dashboards are extremely helpful in enabling decision makers to extract the most intelligence out of the data.

Big data analytics will address critical business questions only when silos of data are broken down—and the resulting data is combined into a composite answer to a business KPI. Unfortunately, many enterprises continue to remain fragmented in their use of information. Finance uses finance systems and marketing uses marketing systems. Big data investments result in improvements only when everyone can exploit the data. Leadership is needed to change organizational thinking on who owns and has access to certain data in order to break down silos and "turf battles." Companies need to create organizational structures where big data crosses silos of expertise within the company to integrate and summarize that data. One strategy is to appoint an overall "champion" or focal point for big data initiatives. This person should be tasked with getting different areas within the company to collaborate on big data metrics or KPIs.

Lastly, given the importance of metrics, it is up to leadership to develop and articulate questions that KPIs need to answer. What exactly do we need to know? An organization needs to look at the right metrics for the phenomena it needs to optimize. These KPIs can then be strategically aligned and agreed upon by all process members with a feedback mechanism that enables continuous improvement.

8.7 The Journey

As companies begin their journey in implementing big data analytics across their supply chains, they need to remember a few key lessons. The first rule is to follow the roadmap just described. The SAM roadmap offers a systematic framework for implementation and integration across the entire supply chain. It helps avoid the hype and falling in the trap of *state of paralysis*.

Remember that there is no one big data solution. Instead, it is a set of analytical techniques for using large data sets that have high volume, velocity, and variety. It is also more than parallel processing or shared service in the cloud. It requires focus, coordination, and a change in culture. Leadership is needed to build organizational capabilities and create the needed infrastructure.

Two additional recommendations will help on the journey: following a maturity map and partnering with a trail guide.

8.7.1 The Maturity Map

Most organizations follow a journey in their implementation effort that builds big data capabilities over time. There appear to be four stages of evolution or maturity along this journey, which are presented as a maturity map (see Figure 8.2). The first and most basic stage is digitizing and structuring the data, called *data structuring*. It consists of the steps that ensure the data is generated, structured, and organized in such a way that it can be used either directly by end users or for further analysis. These techniques include "scrubbing" the data to remove errors and ensure data quality, placing data into standard forms, and adding metadata that describes the data being collected.

The second stage of evolution requires making the data available to all. It can be a powerful driver of value in and of itself, and it can also be an important first step in integrating data sets to create more meaningful business insight.

The third stage is applying basic analytics, which essentially covers a range of methodologies, such as basic data comparisons and correlations. Here, relatively standardized quantitative analyses are used, such as descriptive analytics. These do not require customized analyses to be designed by people with deep analytical skills as in the subsequent stage.

The fourth and highest level is applying advanced analytics, such as predictive analytics, automated algorithms, and real-time data analysis that can create radical new business insight. They allow new levels of experimentation to develop optimal approaches to targeting customers and operations, and opening new big data opportunities with third parties. Leveraging big data at this level often requires the expertise of deep analytical talent.

Figure 8.2 Maturity map of big data implementation

As companies gain competency, they move along the maturity map. It is a mistake for companies to try to jump to stage 4 of the maturity map without first going through the first three stages, which require organizational adaptation and learning.

Leading organizations have discovered that when implementing big data analytics and moving through the maturity map, it is best to begin on a small, targeted, and well-selected pilot project. This is called a "process of purposeful experimentation" and can be the best path toward becoming an organization that fully leverages big data. It also enables learning. This approach is markedly different—and much more effective—than a complete plan for the enterprise prior to doing any implementation. Selecting a few high-potential areas in which to experiment with big data—for example, Web marketing or store layout in retail—and then rapidly scaling successes can be an effective way to begin the journey. It is easier to create value from such small projects rather than jumping directly to complex, analytical big data levers.

An excellent example of this is offered by Kaiser Permanente in California.[29] The company initially concentrated its big data efforts on one IT project exclusively for patients with long-term conditions. The company moved along the maturity map, creating specific disease registries and panel management solutions, rather than an all-encompassing IT solution that addresses a range of problems. This type of slow and targeted approach—particularly as it supports competitive priorities—is the best strategy for starting out.

8.7.2 Partner with a Trail Guide

The effective use of big data analytics in supply chains depends not only upon the availability of data and analytical tools, but also upon the ability of leaders, managers, and employees to use them effectively. Companies have increasingly needed more analytical skills and the gap between available technology and the ability to use it is increasing.

8.7.2a Problem

Analytics is not part of traditional capabilities and core competencies of most companies. To use analytics most effectively, companies are finding that they need to acquire or develop personnel with a variety of analytical skills. First, companies need highly skilled analytical professionals capable of deep data analysis and model development. Second, they need individuals who can conduct analysis in specific functions, such as marketing and merchandising. Third, they need individuals who may not be "quant geeks" but who do understand the basics behind analytics and can communicate with company leaders.

The problem of lack of talent is especially acute in areas where analytics has been new but growing. A good example is web analytics, where skills are difficult to find. Even when skills with individual analytical techniques can be found, it is more difficult to find analysts who are skilled in a variety of techniques. This is a problem as one of the key ways to leverage big data is to be able to integrate a variety of analytical technologies.

8.7.2b Solution

Companies are addressing this problem in a variety of ways. The online flower and gift retailer 1-800-Flowers.com now makes an understanding of analytical business approaches a key criterion for being hired or promoted as a manager.[30] The company also cycles employees from different functions through its Customer Knowledge analytical group so that they can build skills and then apply them in other functions and units when they move back out of the central analytics group. Similarly, department store chain Dillard's has introduced a variety of new analytical tools. To speed adoption of these tools, Dillard's IT function has sponsored training programs and visits by early adopters to try to build understanding and motivation for more rapid general adoption.[31]

The most dramatic new trend is the emergence of analytical outsourcing—which was discussed in Chapter 7, "Impact on 'Buy.'" Most companies lack the capabilities to do all the analytical work that they require. Many need software from external providers, they might need external data, and they likely need someone to help orchestrate the pieces. Many providers are aware of this need and in addition to software and data are now providing analytical assistance.

Earlier chapters offered examples of leading companies that have outsourced various aspects of their analytics capability, from Walmart outsourcing to Mu Sigma to Limited Brands and Pottery Barn working with Alliance Data.[32] Other examples abound. Liz Claiborne uses Fujitsu NSB Group for cross-channel customer management. Goldspeed.com, an online discount jeweler, uses Commercialware for customer service and order management.[33] The lesson here is that companies cannot "go it alone" when it comes to implementing big data analytics and leveraging its full power. Companies need a partner—whether that is a strategic alliance or a transactional outsourcing relationship.

8.7.3 Analytics Outsourcing Strategy

As with any outsourcing engagement, companies need to develop a strategy for which analytical capabilities they want to build for themselves and which they will source from partners. These are important strategic decisions. Companies should also compare providers, considering partners who provide analytical efficiencies for their entire industry versus those who only work with particular companies. The idea here is that it may be possible to gain a competitive advantage by partnering with those outside of their industry. Further, selection of external providers of data and analytics should consider not only the company's immediate need, but also the trajectory of capabilities needed over time. Remember that needed capabilities will only grow as technology advances further. Some partners may build short-term technical capabilities, while others can help to build long-term organizational capability with analytics.

9

Making It Work

Dell, Walmart, and Amazon have focused their analytics efforts on improving supply chain functionality; Harrah's, Capital One, and Neiman Marcus have targeted improvements in customer service and loyalty; Progressive Insurance and Marriott have focused on pricing; Honda and Intel have targeted product and service quality; Novartis and Yahoo! have focused on R&D.[1]

Why the difference? The reason is strategy.

9.1 Strategy Sets the Direction

Leading companies do not just apply analytics randomly where they can. They focus and target their big data analytics efforts on the areas that promise to create the greatest competitive advantage. And this is defined by strategy.

One huge mistake that companies make is to "follow-the-leader" when implementing new processes. We have seen this mistake repeated with any new business process or technology—just-in-time or "lean," Six Sigma processes, Total Quality Management (TQM), and increasingly with "green" initiatives. Adoption of any business initiative needs to support the company's unique business strategy. Not everyone needs the same level of technology or the same degree of implementation. Strategy needs to drive what—and how much—the company needs.

9.1.1 Business Strategy

A company's business strategy provides a plan that clearly defines the company's long-term goals, how the company plans to achieve these goals, and the way the company plans to differentiate itself from its competitors. Strategy sets the direction for the entire company. These decisions are broad in scope and long term in nature. They set the tone for other, more specific decisions. They ask questions such as the following:

- What market are we in?
- What are the unique features of our product?
- How do we compete in the marketplace?

These decisions are important as they define exactly how the company competes in the marketplace and understand what differentiates it from its competitors. For Neiman's, it was customer service; for Progressive, it was pricing; for your company, it may be something entirely different.

Strategic decisions inform tactical decisions, which are narrow in scope and short term in nature. These two sets of decisions are intertwined. Strategic decisions are made first and determine the direction of tactical decisions, which are made more frequently and routinely. It might seem basic, but companies cannot forget to let strategic decisions drive the other, more tactical decisions.

Tactical decisions are specific, short term in nature, and focus on more specific day-to-day issues, such as the quantities and timing of specific resources, and how specific resources are used. They are bound by strategic decisions. Tactical decisions must be aligned with strategic decisions because they are the key to the company's effectiveness in the long run. Tactical decisions provide feedback to strategic decisions, which can be modified accordingly.

Without the direction provided by strategic decisions, companies may pursue competition in areas that do not directly contribute to the business plan and waste resources. This is equally true with analytics implementation. Companies should develop an analytics strategy that directly supports the business strategy.

9.1.2 Analytics Strategy

Organizations need to develop an integrated analytics strategy for the entire enterprise, which supports the business strategy. Data models, architectures, and attributes of solutions need to be considered holistically. Take customer data as an example. A common problem is for individual business units, in silos, to develop their own data strategies without planning to share or aggregate data across the organization. As a result, organizations often discover that they do not even have a common definition of their customers and attributes that apply across the entire organization. Even within the same business unit, such differences can occur. The lack of a customer-centric view severely limits the organization's ability to use any of the powerful big data levers to create new value.

Centralizing and consolidating the IT function can be helpful. The nature of analytical strategy increasingly demands that analytical initiatives and applications be integrated, or at least coordinated. Pricing optimization, for example, affects assortments, logistics, marketing, and financial processes. Centralization is a strategy taken by a number of companies. Best Buy, for example, has created a Project Management Office (PMO) for analytics to coordinate the company's various analytical initiatives.[2] The PMO has a focus on cross-functional initiatives. Similarly, Starbucks, the coffee chain, has centralized most of its analytical activities in the corporate strategy organization.[3]

In addition to the imperative of business strategy, companies need to consider a few additional aspects of their analytics capabilities when developing an analytics strategy: technology gap, outsourcing strategy, analytics supply chain strategy, and short- versus long-term strategy.

9.1.2a Technology Gap

An effective analytics strategy must consider the issues of data architectures, data models, and attributes of solutions in unison, including security and compliance, and frontline services. Many organizations will require additional investment in IT hardware, software, and various IT services to achieve the capability levels they desire. The level of investment will vary considerably depending on a company's current state of IT capability.

Companies need to conduct a gap analysis to assess and identify gaps in the technology their enterprise currently has versus their current and projected needs. This includes technology for effectively capturing, storing, aggregating, communicating, and analyzing data. Many companies will find that they have legacy systems that need updating. The level of technological investment can be enormous. Companies should not play "follow the leader" copying others in their industry. Rather, they must define their analytics strategy in unison with the business strategy and the technological leadership role of the organization. Business leaders then need to develop a business case for new investments and prioritize that spending.

9.1.2b Outsourcing Strategy

An important part of the analytics strategy is to identify which capabilities companies want to build for themselves and which they will outsource to partners. This is an important outsourcing decision. It is also a strategic decision. While companies cannot go it alone on the IT quest given the rapidly advancing state of technology and the capabilities required, outsourcing can also lead to dependencies. Companies may choose to keep certain capabilities in-house as well as maintain control over certain data sets.[4]

A part of this strategy is deciding on choice of partners. One strategy may be for a company to consider a partner that provides services for an entire industry, thereby creating analytical efficiencies. An example is Catalina Marketing, which manages loyalty programs for more than 200 grocery chains in the United States and some in Europe.[5] Another strategy is to partner with a company that is not industry specific, thereby providing a competitive advantage. An example is dunnhumby, a provider of customer analytics service. dunnhumby works with retailers such as Kroger and Tesco. However, its clients cut across geographies and industry segments.[6]

9.1.2c Analytics Supply Chain Strategy

Companies need to consider their analytical supply chain strategy, as they integrate and share data with their supply chain partners. They need to ensure that they are able to combine external data sources with their own transaction data so it can be analyzed in an integrated approach. Data aggregation and sharing are needed for integration. However, companies need to carefully consider which partners could benefit from access to their data and analyses, and selectively make them available.

Access to data is obviously beneficial from an information-sharing standpoint, but does raise risks. Just consider the recent incident when the retailer Target had a security data breach where up to 110 million customers had their personal data stolen. This included credit and debit card data, customer names, and PIN numbers. Although Target's CEO Gregg Steinhafel said the company was "truly sorry" for the data breach, Target came under sharp criticism and saw a sharp drop in sales and stock price.[7] According to New York Attorney General Eric Schneiderman, "Consumers expect and deserve companies that protect their personal information when they shop on their websites and in their stores."[8] Companies need to be extraordinarily careful that, through the eagerness of quickly adopting data-capturing technologies, they do not compromise themselves or their supply chain partners.

9.1.2d Short Versus Long Term

Any decisions about collaboration with external providers of data and analytics should consider not only the retailer's immediate need, but also the trajectory of capabilities needed over time. Some partners may build short-term technical capabilities; others can help to build long-term organizational capability with analytics. Given the high costs of technical requirements, it is often tempting for companies to go with a short-term solution.

Remember, however, that the need for this capability will only continue to accelerate and that technological capability is exponentially growing. This means that short-term investments made today without a plan will quickly become long-term legacy systems of tomorrow (see Figure 9.1).

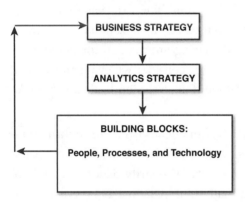

Figure 9.1 Strategy determines building blocks

9.2 The Building Blocks

Three building blocks define the foundation of a company's analytical strategy. The building blocks are people, processes, and technology. The characteristics of these building blocks will vary based on the company's business strategy and analytics strategy. Defined together, these building blocks need to support the company's analytics strategy. The following sections look at these three building blocks (see Figure 9.2).

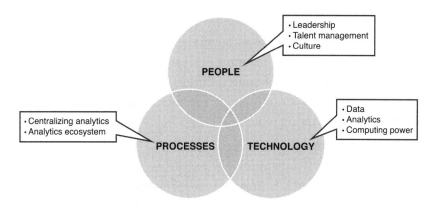

Figure 9.2 The building blocks (a mix of people, processes, and technology)

9.2.1 People

Here we discuss the role of People, the first building block. This consists of leadership, talent management, and culture.

9.2.1a Leadership

Big data analytics does not eliminate the need for leadership. In fact, the success of big data analytics cannot happen in organizations without leadership. It is leadership that provides the vision, the strategy, the goals, and the path to succeed. It is leaders who are able to see great opportunities, understand how markets develop, and have the insight to propose novel offerings.

Leaders are the ones who develop the strategy and articulate a compelling vision. They are also the ones who create the culture needed to support any organizational change or process implementation. It is leadership that ensures the organization is moving in the right direction. It is also leadership that ensures resources are allocated for the organization to accomplish its goals. This includes investments in information technology, talent recruitment, training and retention, and organizational incentives to create a data-driven culture.

Competing and capturing value using big data requires leaders to address specific barriers across talent—technology, talent, and processes. Successful analytics companies have leaders who can do all these things. The other key prerequisite is leadership that creates an interest—and even passion—about analytics. Leaders need to emulate what they are trying to motivate.

It is up to leadership to make sure that the right capabilities are in place and that the organizational incentives, structures, and workflows are aligned. The UK retailer Tesco, for example, has developed a strong, data-driven mind-set from top leadership to the front line. The company has integrated customer intelligence into its operations at all levels from a variety of consumer-targeted big data strategies.[9] At Famous Footwear, the executive team meets with the head of testing every two weeks to discuss results and plan data-gathering and evaluation programs. At Amazon.com, it is said that Jeff Bezos fired a group of Web designers who changed the company Web site without conducting experiments to determine the effects on customer behavior.[10] It is this kind of leadership that ensures that employees at all levels leverage insights derived from big data.

9.2.1b Talent Management

Companies need to establish a culture that values and rewards the use of big data in decision making. People respond to incentives—both intrinsic and extrinsic. Leaders can talk about initiatives, but it is incentives that show commitment. Companies also need to recruit and retain deep analytical talent and retrain their analyst and management ranks to become more data savvy.

- **Recruiting Talent**—Leading big data companies have built a critical mass of deep analytical talent upon which the rest of their organization can draw. These companies start with early hires that make up the core around which they build their analytics teams. These initial hires need to be the most capable people in order to build the most effective teams possible. Given the potential competition for this talent, however, organizations must be aggressively recruiting deep analytical talent.[11] This may involve sourcing talent from other geographical areas, outsourcing select analytical services from vendors, or aligning with an academic institution to get access to newly minted graduates.

- **Managing Talent**—Having a core set of deep analytical talent is not enough to transform an organization. Leadership has to have an appreciation and understanding of analytics to drive the right culture. Otherwise, leaders and analysts do not know how to take advantage of big data capability. Leaders in an organization need to develop at least a rudimentary understanding of analytical techniques to become effective users of these types of analyses. Organizations need to develop training programs to increase analytics understanding and capabilities of everyone in the organization—including leaders and managers. To take advantage of big data analytics, organizations need to change the way that they use it and understand how to use it.[12] Consider Capital One, the financial services firm. The company has created an internal training institute called Capital One University. It offers professional training programs, such as training on testing and experiment design.[13]

9.2.1c Culture

Organizational culture is composed of the values, behaviors, and unwritten rules that contribute to the social and psychological environment of an organization. Analytics-driven companies have a companywide culture that has a respect for measuring, testing, and evaluating quantitative evidence. It is a culture that promotes basing decisions on hard facts at all management levels. This includes top leadership as well as lower-level employees. Senior executives set an example with their own behavior, exhibiting confidence in facts and analysis. Decisions are made based on supporting evidence rather than mere opinions. It is part of the accepted behavior.

Employees also know that their performance is gauged the same way—a fact that can, in and of itself, raise performance. Everyone behaves accordingly. In these organizations, even human resource decisions are based on metrics—such as rewarding compensation. A report on Harrah's, for example, depicts a significant change in its rewards culture.[14] The culture was previously based on paternalism and tenure. This was replaced by a culture where performance measurements are based on data and facts, such as financial and customer service results.

An analytics culture can sometimes have tension between innovative or entrepreneurial impulses and the requirement for evidence. These are hard decisions that blend strategy and culture. Just consider how many years it took for 3M to develop sticky notes. Still the amount of tolerance for blue-sky development where designers or engineers chase after ideas varies. In many analytics-driven organizations, R&D, like other functions, is rigorously metric driven. At companies such as Yahoo!, Progressive, and Capital One, process and product changes are first tested on a small scale.[15] The changes must be numerically validated before they are implemented on a broader scale.

9.2.2 Processes

The Processes building block includes the potential to centralize analytics and to develop an analytical ecosystem.

9.2.2a Centralizing Analytics

Analytics has traditionally been siloed within particular functions and units within organizations. It has often been a backroom operation or support function. As companies increasingly become analytically driven, they are moving toward centralizing their analytical resources. The purpose is for analytics to serve the entire enterprise. The resources include data, people, and technology. They are not only available to everyone in the organization, but they are also intertwined with organizational processes.

In most organizations, analytics or IT resides in a particular function, area, or department. This may be logistics, operations, or marketing, or it may be a separate isolated entity. Typically, these functions have their own data, analytical people, and software tools. However, competitiveness requires the development of an analytics strategy that is integrated throughout the organization. Analytics needs to coordinate these efforts. To create and manage integration, many leading companies are moving toward centralizing analytics groups. Centralization makes analytics available to everyone. Centralized analytics groups can coordinate the management of all applications, enterprise data warehouses, and analytical software. They may also coordinate analytical efforts that are done more locally, such as at the store level, closer to the customer.

Walmart has several different analytics groups, but is increasingly coordinating and centralizing analytical activity.[16] Sears Holdings has created a central group of analysts and Home Depot has two major centralized groups to coordinate analytical work across functions and stores.[17]

Companies that are serious about big data analytics should begin to network and coordinate analytical activities across the enterprise. Companies that want to be on the cutting edge of analytics should probably establish a central function to manage and facilitate analytics somewhere in their organizations. Centralization will ensure that everyone has access to the data, that there is integration, and that there is no duplication. Further, this will ease implementing an analytics strategy. There need to be mechanisms in place to ensure that analysts work on projects that support the business strategy, that address the core needs of the business, and that the analytics activity remains close to business decision makers.

9.2.2b Analytical Ecosystem

An important part of using big data analytics across the supply chain is structuring collaborations with external partners to increase analytical maturity. As you have seen through countless examples, succeeding with analytics cannot be done alone. Instead, companies need to collaborate with multiple partners in an *analytical ecosystem.*[18]

This ecosystem will be composed of various value-adding partners. It may include suppliers, channel partners, external providers of data and analytical services, and software and hardware providers.

Most companies lack the capabilities to do all the analytical work that is required. They also cannot keep up with the pace of technological advancement. This is just not their core competency. They will almost certainly need various capabilities from external providers, including software, applications, and external data.[19] They may need training and talent management.

As companies develop their analytics strategy, they need to decide which capabilities they will outsource and which skills they will keep internally. Also, they will need to decide who will make up the core analytics partners. Some external providers are exclusive to a particular company. Others, however, make their products and services available to the entire industries or segments. Recall that software firms SAS and Teradata work with companies not only to implement their solution offerings, but also to help solve particular analytical problems and create applications. Accenture provides consulting and outsourcing services to a variety of companies, including Best Buy.[20]

9.2.3 Technology

Of course, we would not be here without Technology. We review here key advances in technology that have enabled big data analytics to get off the ground.

9.2.3a Data

The key element of technology is data. A company cannot do analytics without clean, high-quality, integrated, and accessible data. Today, companies increasingly have vast amounts of it available— from point-of-sale (POS) transactions, from Web sites, from credit programs, from current loyalty programs, from RFID and other sensors, and from other business applications. However, this data must be "scrubbed" and cleaned. It must also be digitized, structured, and organized in such a way that it can be used either directly by end users or for further analysis. This process ensures data quality, placing data into standard forms, and adding metadata that describes the data being collected.

Data must be accessible to be analyzed. The data must be visually presentable for decision makers to be able to integrate it into decision making. Much work has been done in creating good data visualization and dashboards. These enable managers to not only view the critical variables, but also be able to manipulate values of certain variables and "see" changes in other dependent variables. These types of tools are very powerful. Without them, it can be easy to become overwhelmed with the volume of data.

9.2.3b Analytics

Analytics is applying statistics and quantitative applications to big data. Many of these applications—such as correlations and regression—have been around for decades. It is big data and computing power that are now able to take greatest advantage of these capabilities. Recall that Google's director of research, Peter Norvig, explained it well by saying: "We don't have better algorithms. We just have more data."[21]

There are two basic categories of analytics. The first is *descriptive analytics*. It is the simpler category of the two. Descriptive analytics is essentially used to summarize and describe a set of data. It tells us *what happened*. The majority of business analytics is descriptive. These are statistics that describe events—such as counting the number of times something happened. This may be how many products were returned, how many calls were received at a call center, how many people are on Expedia searching for flights from Miami to Las Vegas, or how many people are on the Google search engine looking for flu symptoms. These numbers can be summarized, aggregated, and correlated. For example, the highest number of people searching for remedies for flu symptoms may be in Oklahoma City, which may indicate the center of the epidemic. From a statistical standpoint, these computations are very simple. It is today's volume and computing power that can give us interesting and unique insights.

Predictive analytics, on the other hand, is much more sophisticated. It tries to tell us *what will happen*. Predictive analytics utilizes a variety of statistical methods, modeling, and data mining techniques to study recent and historical data. It then applies sophisticated algorithms that enable analysts to make predictions about the future. Realistically, predictive analytics cannot "foresee" the future. No one can. Predictive analytics is simply sophisticated forecasting tools. It gives us a probability—or a likelihood—that something will happen.

This is exactly what IBM's Watson computer is doing in a project with Memorial Sloan Kettering Cancer Center (MSK) in New York City.[22] Watson is being used to sift through volumes of data on patients and treatments and recommend the best course of treatment for a particular patient. Watson uses predictive analytics and does not provide an answer—or tell exactly what the future holds. Rather, Watson provides choices in courses of options. For example, for a particular cancer patient—considering all available patient data—Watson may recommend three courses of action. One may be with a 95 percent confidence level, another with a 45 percent confidence level, and the third with a 10 percent confidence level. Watson then offers justifications for its recommendations. Physicians can then look at this information and make the final decision. This is really what predictive analytics does. It is not a crystal ball but simply gives probabilities of outcomes.

9.2.3c Computing Power

The volumes of big data required for analytics applications are typically well beyond the capacity of low-end computers and servers. As a result, many analytics companies are moving to technologies designed to process large amounts of data quickly. This is made possible with today's massive computing power, which is available at a lower cost.

One example of such technologies is Hadoop, an open source platform. It is specifically designed to deal with big data that is a mixture of complex and structured data that does not fit nicely into tables. Hadoop uses distributed applications across many servers so that the database is actually distributed over a large number of machines. Spreading data over multiple machines greatly improves computing capability, reduces index size, and substantially improves search performance. The database is typically divided into data partitions called *database shards*. A database shard can be placed on separate hardware and multiple shards can be placed on multiple machines. Database shards significantly improve performance. They can also be based on real-world segmentation of the data that can help analysis. For example, you can compare customers across distribution channels or demographics. This makes it especially easy to query a particular segment of the data or evaluate comparisons across segments.

Large data—coupled with larger and affordable computing power—enables companies to do on a larger scale what cannot be done on a smaller one. Computers have become faster, memory has become more abundant, and storage space has expanded through cloud computing. This high capability coupled with low cost has leveled the playing field for small and medium-sized companies. All these companies can now become analytics competitors.

9.3 Following the Maturity Map

Organizations follow a journey as they build their big data analytics capabilities over time. Chapter 8, "The Roadmap," introduced the maturity map, which shows the stages of evolution along this journey.

The first stage is data structuring. It is the most basic stage that involves digitizing and structuring the data. It consists of the steps that ensure the data is generated, structured, and organized in such a way that it can be used either directly by end users or for further analysis. These techniques include scrubbing the data to remove errors and ensure data quality, placing data into standard forms, and adding metadata that describes the data being collected. Companies are investing many millions of dollars in systems that pull in data from every source possible. Enterprise Resource Planning, customer relationship management, point-of-sale, and other systems ensure that no transaction or other significant exchange occurs without it being gathered. But to compete on that information, companies must present it in standard formats, integrate it, store it in a data warehouse, and make it easily accessible to anyone and everyone. And they will need a lot of it. For example, a company may spend several years accumulating data on different marketing approaches before it has gathered enough to reliably analyze the effectiveness of an advertising campaign. Dell employed DDB Matrix to create a database that includes 1.5 million records on all the computer maker's print, radio, network TV and cable ads, and sales for each region.[23] That database took a period of more than seven years to be created, providing information that enables Dell to fine-tune its promotions for every medium in every region.

The second stage of evolution requires data availability. To compete on that information, companies need to integrate the data, store it in a data warehouse, and make it easily accessible to everyone. The third stage is applying basic analytics, which covers a range of methodologies, such as basic data comparisons and correlations. At this stage, descriptive analytics tends to be used, which involves relatively standardized quantitative analyses.

The fourth and highest level is applying advanced analytics. In addition to descriptive analytics, which explains what has happened, here, companies use predictive analytics. They can use automated algorithms for monitoring and prediction. They also conduct real-time data analysis that often can create radical new business insight and models. At this new level, experimentation is conducted to develop optimal approaches to targeting customers and opening new big data opportunities. Leveraging big data at this level requires the expertise of deep analytical talent (see Figure 9.3).

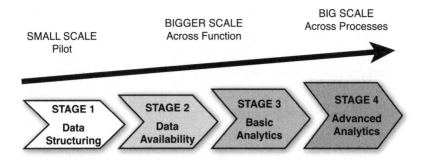

Figure 9.3 The maturity map

As companies gain competency, they move along the maturity map. It is a mistake for companies to try to jump to stage 4 of the maturity map without first going through the first three stages, which require organizational adaptation and learning.

9.3.1 From Small to Big

The transition to using big data analytics should proceed cautiously with a pilot project before scaling up to firm-wide integration.

9.3.1a Select the Pilot

Leading organizations have discovered that when implementing big data analytics and moving through the maturity map, it is best to begin on a small, targeted, and well-selected pilot project. This is called a *process of purposeful experimentation* and can be the best path toward becoming an organization that fully leverages big data. This approach is markedly different—and much more effective—than a complete plan for the enterprise prior to doing any implementation. Selecting a few high-potential areas in which to experiment with big data—for example, Web marketing or store layout in retail—and then rapidly scaling successes is the most effective way to begin the journey. It is easier to create value from such small projects rather than jumping directly to complex analytical big data levers. Targeted projects help a company to learn what works and to begin developing capabilities. A good example is Kaiser Permanente in California.[24] The company initially concentrated its analytics efforts exclusively on patients with long-term conditions by creating specific disease registries and panel management solutions. This focused approach led to much faster time to impact rather than an all-encompassing IT solution that addressed a range of problems.

It is important that the pilot project selected is not random. Rather, it should be driven by strategy and the understanding of what competitive priorities are important to the organization. Even though it is a pilot project, resources will nevertheless be allocated. This is the reason why companies like Walmart first focused on the supply chain functionality, whereas Neiman's focused on customer loyalty. The SAM roadmap discussed in Chapter 8 shows that best practices start with strategy, then align all participants, and then measure. This is true whether working on a small or large scale.

This process also enables learning. For example, companies such as Progressive and Capital One first test and implement process and product changes on a small scale.[25] Once these are validated, they move to broader implementation. This approach is well established in both academic and business circles and should be the process followed.

9.3.1b Then Move Bigger

To be effective, the implementation is moved to the broader function and then across processes of the organization and the supply chain. Organizations, and entire supply chains, can be viewed as a collection of processes, rather than just a collection of departments or functions. For the greatest impact, it is important to implement across processes. A *business process* is simply a structured set of activities or steps with specified outcomes. The sequence of process steps goes beyond the organization and cuts across a supply chain network.

For example, there is the customer service process, which involves a series of activities designed to enhance the level of customer satisfaction by meeting or exceeding customer expectations. This may involve a series of well-coordinated activities, such as billing and invoicing, handling product returns, providing real-time information on promised shipping dates, and product availability. A company may pilot its initial effort along one activity such as handling product returns. For full effectiveness, however, this effort needs to be expanded across the entire customer service process.

Another example is the order fulfillment process, which involves ensuring that customer orders are filled. It may involve activities such as receiving and processing the order, ensuring movement of product and delivery, and engaging in customer follow-up. Other examples of processes include the manufacturing process, which involves ensuring production of products; the demand management process, which balances demand requirements with operational and supply chain capabilities; and the distribution process, which involves distributing and delivering products to specified locations. Organizations have many other processes, each having a series of activities designed to create a particular output for the customer, whether it be a service or a tangible product. The output is a result of the process that produces it. To improve the output, a company must improve the process.

Processes cut across many organizational functions as shown. For example, the customer service process cuts across a number of different functions. It must involve marketing that interfaces with the customer, logistics that ensures product delivery and movement, and operations that may deal with repairs. Similarly, order fulfillment requires operations to ensure order availability and processing, logistics to arrange for order picking and shipping, and marketing for customer follow-up. For processes to be effective and efficient, organizational functions must work together and be well coordinated. Sales & Operations Planning (S&OP) is an excellent place to achieve this type of implementation as S&OP requires functional coordination that is dependent upon data (see Figure 9.4).

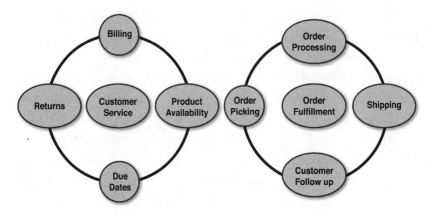

Figure 9.4 Organizational processes cut across functions

9.4 Sales & Operations Planning (S&OP)

Sales & Operations Planning (S&OP) has been a popular and much-discussed concept among business leaders. It is a highly effective business process that relies on cross-functional integration and data-driven decisions. In fact, data is the basis of S&OP. Accurate, clean, and reliable data is paramount to its success. Therefore, integrating big data analytics into this process is an important implementation. Following key rules for implementation is imperative to achieving success.

9.4.1 What Is S&OP?

S&OP is an integrated business management process intended to match supply and demand through functional collaboration. It is a process through which the executive leadership team achieves strategic focus, functional alignment, and synchronization across the entire organization. Although conceptually this sounds simple, one of the key attributes is that it is an *ongoing, proactive, intracompany process.* It is a process that requires ongoing communication and continuous improvement. Done correctly, the process allows executive management to anticipate business changes without resorting to late, reactive, and costly responses. Without this coordinated effort, companies are typically plagued with short planning horizons, lack of functional alignment, and reactive responsiveness.

The objective of S&OP is for the functional organization to reach a monthly consensus regarding a single operating plan. Consequently, the multistep S&OP process accepts inputs at different stages from sales, marketing, product development, operations, sourcing, finance, and logistics. The resulting plan allocates the critical resources of raw materials, people, capacity, time, and money to most effectively meet market demands in a way that maximizes profits for the company. This monthly plan must be reviewed and approved by executive management as the final step. In this way, the senior leadership ensures that the plan focuses on the best interests of the company across the entire planning horizon as well as meets strategic objectives and financial plans.

S&OP is needed to ensure that the tactical plans of each functional organization align with the company's strategic business plan—a challenge for many companies. It forces all functions to base their individual plans on the same single set of numbers or "one truth." It also promotes teamwork between the different functions as it encourages them to work together to achieve the company's strategic goals. The recurring, monthly nature of the S&OP process is also needed to drive continuous improvement across all functional areas.

S&OP provides numerous benefits—both tangible ("hard") and intangible ("soft"), as shown in Table 9.1. The former includes faster and more controlled new product launches, more stable production and service rates, shorter customer lead times, and higher levels of customer service. In addition, S&OP lowers inventory levels, thus reducing both inventory carrying costs and safety stock costs. However, it is the intangible benefits of the S&OP process that are readily apparent in a company's culture and the attitude of its employees. This includes higher levels of trust and openness between members of different functional organizations; enhanced teamwork between executive management and middle management; and the making of better, more accurate decisions, quickly and with less effort.

Table 9.1 S&OP Delivers Hard and Soft Benefits

"Hard" Benefits	"Soft" Benefits
Improved customer service	Improved teamwork
On-time deliveries	Tight linkage between strategic plans and day-to-day activities
Reduced inventory levels	More effective communications
Improved inventory turns	Better informed decision making
Improved plant efficiency	More advanced financial plans
Less plant downtime	Greater accountability
Reduced logistical costs	Larger amount of control
	A view into the future

9.4.2 The S&OP Process

The S&OP process is conceptually simple to understand. A properly working S&OP process aligns the company's supply and decision making with the customer's product demand and desired service levels. This occurs on a monthly basis that typically covers a forward-looking 12-month to 24-month planning horizon. This process is composed of five distinct steps: generation of month-end reports, demand planning, supply planning, alignment meeting, and executive S&OP meeting. It is important to remember that these steps make up a cycle that is continuous in nature. Step 5, which occurs at the end of the cycle, leads to step 1, which is the beginning of the new cycle. Mistakes from the previous cycle serve as lessons for the next cycle following continuous improvement principles (see Figure 9.5).

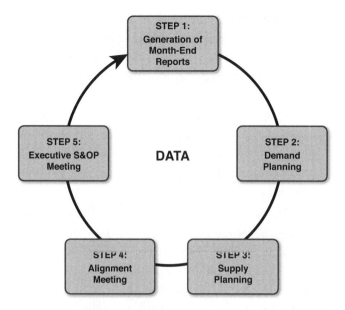

Figure 9.5 Monthly S&OP cycle

There are five steps to the monthly S&OP cycle:

- **Step 1**—The monthly S&OP process begins with the generation of month-end reports. The data is formatted and prepared
  for the demand planning team. In addition, production and
  inventory data is sent to operations to be used as part of their
  evaluation of manufacturing capacity. Having clean, reliable,
  and trustworthy data is essential.

- **Step 2**—The next step is demand planning, which is the
  responsibility of sales and marketing with support from product development. The goal is to generate a quantitative demand
  forecast indicating expected demand, sales, and the resulting
  expected revenue. Statistical analytics in conjunction with an
  assessment of current and future business conditions aids in
  developing this forecast. This single demand plan is passed on
  to the supply planning process team.

- **Step 3**—The supply planning process is the responsibility of
  operations, including purchasing, materials planning, and manufacturing. This determines if the company has the necessary

inventory and manufacturing capacity to meet the forecasted demand plan. In addition, financials are evaluated for material and human resources necessary to meet the required capacity. The supply planning team proposes adjustments to the demand plan where it is necessary or practical.

- **Step 4**—This is a joint preexecutive management meeting to review the adjusted demand forecast and resolve any outstanding resource or capacity issues. This meeting must include sales, marketing, product development, operations, sourcing, finance, and logistics.

- **Step 5**—The fifth and final step is the executive S&OP meeting. This is a high-level, fact-driven review with the goal of finalizing the monthly demand forecast and the associated operations plan. In addition, it gives executive management the opportunity to modify these two monthly plans to fill identified gaps between future projections and the company's strategic objectives as a result of changes in the internal and external business environments.

9.4.3 Data—The Basis for "The Conversation"

S&OP cannot happen without data. The process crosses organizational boundaries where internal tensions between departments may create barriers to implementation. The process relies on having reliable, clean, and accurate data. This is the key to success. Trustworthy data serves as a basis for "the conversation." The S&OP process requires data to be aggregated from multiple functional organizations. However, to move the project along, data must provide useful information, with the least amount of data possible. Otherwise, the project is held up by additional data-gathering efforts. Consequently, two significant decisions vital to implementation are determining *what data* needs to be gathered and *how* that data should be aggregated. Once made, the integrity, quality, and reliability of the data must be guaranteed on a monthly basis.

The final step regarding data management is ensuring that executive management receives the information in an easily understood format. This facilitates quick, well-informed decisions during the executive S&OP meeting as well as identification of any potential gaps that need to be addressed. The key here is to clearly understand and articulate the problem, information needed, and visualization of data and facts.

9.4.3a Phased Implementation

The old adage of crawling before walking applies here. Companies cannot jump into full S&OP integration but must go through a phased implementation approach. In the first phase, although there is communication between functions, the decision-making process is effectively siloed. This is where pilot projects can take place, can be tested, and can be validated. This process then moves into a cross-functional, decision-making phase that is reactive and designed to address problems. Data sharing is critical here to move the organization to the next step. The third phase moves cross-functional decision making from reactive to proactive, which is forward looking. In the last and final stage, the organization is fully integrated both horizontally across the organization, and even beyond the enterprise, as well as vertically, integrating strategic and tactical decisions (see Figure 9.6).

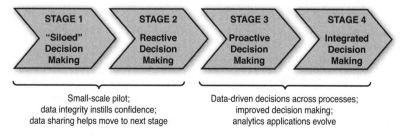

Figure 9.6 Stages of S&OP implementation

9.4.4 Measurement and Continuous Improvement

Performance measurement is vital to both the S&OP process and big data implementation. It is a key element of the SAM roadmap. This can be accomplished using a balanced scorecard, strategically aligned KPIs agreed upon by all process members, and a feedback mechanism that enables continuous process improvement. Implementation must include development of a scorecard that aligns key performance indicators (KPIs) with strategic goals. These KPIs must measure performance at all levels: strategic, tactical, and operational. Examples include increasing profit margins (strategic), lowering inventory levels (tactical), improving on-time delivery (tactical), and increasing daily product yield (operational). It is critical that this scorecard balance the perspectives of the different functional organizations with customer expectations and strategic company goals.

It is also important to select metrics that help make tradeoffs. One such metric is *working capital,* which requires closely monitoring inventory levels at all phases and making tradeoffs. For example, senior management throughout the S&OP process may be forced to make sacrifices to service levels to select customers in order to maintain a level of inventory that keeps working capital at the optimal level. This requires a hard look at safety stock, work in progress, and pipeline stock. Ultimately, management is striving for the highest level of customer service with the lowest working capital investment. This could lead to choosing different forms of transportation, shipping from different locations, or simply choosing to not meet certain customer requirements.

Finally, the company must strive for continuous improvement by setting higher KPI levels when current levels are achieved and comparing performance against benchmarks from other companies in the industry segment. Only in this way can the company know if it is truly benefitting from the S&OP process and big data analytics efforts.

9.5 People Making Decisions with Data

It is ultimately people—whether individuals or teams—who make decisions based on big data analytics. As human beings, we have limited ability to consume and understand data. This has been documented in hundreds of studies and is an academic discipline all in itself.[26] Big data techniques can greatly assist human decision making. Recall IBM's Watson providing doctors with alternative treatments with probabilities of outcomes—but leaving the final decision up to the physicians.

There are ways to help individuals and teams process, visualize, and synthesize information from big data. Teams, such as S&OP implementation, need to understand how to use the data for decision making, recognize which data is important, and be able to gain intelligence from the available information. For instance, more sophisticated visualization techniques, algorithms, and dashboards can enable people to see patterns in large data sets, which can help reveal the most pertinent insights. S&OP decisions are made in teams with the use of visualization of technology. Getting this right provides tremendous opportunities and has the potential to improve efficiency and effectiveness, enabling organizations both to do more with less and to produce higher-quality outputs. Incorporating such technologies to improve decision making—in S&OP and all big data implementation efforts—will go a long way to turning big data information into intelligence.

10

Leading Organizational Change

Here is the latest from Amazon. The company is currently working on a plan that would ship products to customers before purchases are even made.[1] The company recently gained a patent for *anticipatory shipping*, a system that would allow Amazon to send items to shipping hubs in areas where it believes these items will sell. Amazon is drawing on its massive stores of customer data. This includes previous searches and purchases, wish lists, and also how long the user's cursor hovers over an item online. Based on advanced analytical algorithms, Amazon wants to engage in *predictive purchasing*, determining where products will go.

Amazon then plans to preemptively box and ship products it expects customers to buy. The company may even load products onto trucks and have them sent to shipping hubs in areas where the company believes the purchases will take place. Obviously, the model optimizes site selection of shipping hubs and quantities, as well as product categories. The potential for this new idea is to cut delivery times down and put the online vendor ahead of its real-world counterparts. Although the scenario may lead to unwanted deliveries and even returns, the company believes these will be relatively small given the increased sales and delivery benefits.

Is this the future? Time will tell whether real-world costs of inventory movement is offset by the gains. Nevertheless, it does show the potential power of big data analytics. It also shows that ignoring big data analytics for supply chain management is at your own peril. Embracing requires leading organizational change.

10.1 Transformation Required

The majority of companies are not Amazon. Most are just embarking on their big data analytics journey. For most, it will take a bit longer to transform themselves into a precision analytical machine. However, given the current supply chain environment, the overall trends are clear. Companies are rapidly becoming data-intensive. They are taking advantage of the large volumes of data to better operate and manage their businesses. The good news is that most companies have only scratched the surface of what is possible.

Big data analytics has demonstrated its potential. However, there are many barriers to implementation. As mentioned previously, the barriers to implementation fall into three general categories: technology, people, and processes. These barriers work together and must be overcome as such. Companies sometimes make the mistake of investing in one, for example technology, with the exclusion of the others. These three barriers need to be addressed simultaneously and their capabilities need to be matched to achieve synergies. After all, improving technology will not create gains if people cannot use it. Companies need to invest in analytics following a plan that specifies changes to all three areas and addresses unique needs of the business, rather than following the hype.

Chapter 8, "The Roadmap," presented the SAM roadmap that outlines an organizational path of implementing big data analytics. The implementation of the roadmap cannot happen without instituting organizational change. Big data analytics—although exciting and full of possibilities—has created a rapidly changing environment that requires adaptation and continuous improvement. For companies to survive and thrive, they must adapt.[2] Implementing the roadmap requires organizational change.

The following sections show you how to lead such a transformation.

10.1.1 Organizational Change

Organizational change is any substantive modification to an aspect of an organization. It may consist of a change in the workforce, technologies, structure, work processes, culture, values, or strategic vision. Yet change is difficult. Half to two thirds of these changes fail to achieve their expected results.[3] In a rapidly changing environment, the knowledge that is most useful to organizations is that which helps them change and adapt.[4] Leaders must understand when change is needed and how to guide their organizations through it. The SAM roadmap details how to implement big data analytics in an organization and across the supply chain. This implementation requires a major organizational change to become a data-driven organization. This section explains how to guide this change process.

The way that change is implemented throughout the organization depends on the type of change involved. There are two types of organizational change. Change can be either transformational or incremental.[5] *Transformational change* occurs when an organization changes its strategic direction or reengineers its culture or operations in response to dramatic change, such as a technological breakthrough or a merger.[6] Transformational changes are typically a reaction to a major problem or challenge in the competitive environment. These changes require leaders and all members of the organization to embrace learning and innovation in order to succeed.[7] They represent a major shift for the organization.

Incremental change occurs when an organization takes relatively smaller steps toward its goal. This may involve restructuring to be more efficient or expanding its line of products or services to promote more growth. These changes occur incrementally or gradually. The organization exists in a state of "equilibrium." The focus with incremental change is on fine-tuning existing practices.

A great example of transformational change in action is what happened at Rakuten Inc.—Japan's online shopping equivalent of Amazon. In 2012, the founder and CEO of Rakuten, Hiroshi Mikitani, made a transformational change to his large company when he instituted an English-only policy for all company communications. This change required that all meetings, training sessions, and even e-mails would be in English.[8] His goal was to globalize the firm, broaden the mind-set of current employees, and move away from the Japan-centric thinking of his employees. Mikitani's actions were dramatic but necessary if the company was to become a global competitor. His actions were intended to lead his organization through a transformational change.

As happens with most changes of this magnitude, his requests were met with a bit of resistance. Some employees resigned in protest while others struggled to meet the new expectations. Originally, Mikitani expected employees to adapt to the change without the organization's help. However, after feedback came in regarding employees experiencing apprehension and anxiety, the company began offering free English classes and time on the job to study. Mikitani has made it clear that speaking English is a requirement for all jobs. The organization is leading transformational change and it is working. Currently, more than half of the company's Japanese employees use English for all internal communications.[9]

The type of transformational change that has taken place at Rakuten is the kind of change required to implement big data analytics and create a data-driven, decision-making culture. Learning to make decisions based on data and information, rather than guesswork, is akin to forcing an organization to communicate in an entirely different language. For this to be successful, there are certain steps that need to be followed in the change process.

10.1.2 How Organizations Change

There are predictable stages involved in organizational change.

10.1.2a Stages in Transformational Change

Transformational change is of a very large magnitude. It requires a major shift in the organization, as you saw in the example of Rakuten.

The process of organizational change has been outlined in the seminal work of Kurt Lewin.[10] Lewin outlined four steps that organizations go through in the change process—if the transformational change is going to occur and last. These are steps every organization goes through. Although approaches differ in the descriptions of how to perform each step, there is consensus on the need for all four steps.

The steps represent a continuum. They are recognizing a need, unfreezing, changing, and refreezing (see Figure 10.1). The unfreezing and refreezing steps are especially important, as for meaningful change to occur, it cannot just be forced on the organization. The organization must be prepared through the unfreezing stage—where old habits "thaw." After change is implemented, it must be solidified—in the "refreezing" stage. Otherwise, the organization will fall back into its old habits.

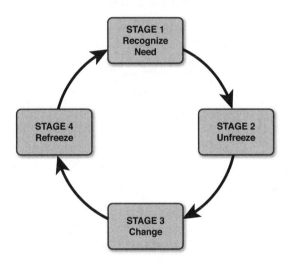

Figure 10.1 Steps in the change process

10.1.3 Top Down or Bottom Up

The conventional perspective to change and learning is that it is essentially a top-down process. It is driven and managed from the top down where leaders work to ensure that their changes are implemented within the organization. The expertise and ideas primarily come from executives or consultants. They provide their expertise and vision related to the organization's environment that needs to be addressed through change. A vision or description of what needs to change is formulated and those at the top persuade others of the need to change, direct, and oversee the changes, and they implement structural and cultural changes to reinforce the new vision.

A different and more sustainable approach to change is more of a bottom-up process. In this process, everyone participates in developing and implementing the change. This process is part of the Japanese style of management seen at companies such as Toyota, Sony, Nissan, and many others. It is also part of the servant leadership orientation that espouses that organizational members are inherently valuable and should be engaged in shaping change. Research comparing top-down leadership with participative approaches shows that when leading change, participative approaches are much more effective. Participation and open dialogue among members are found to be essential in gaining a commitment to change.[11]

Do not misunderstand. The bottom-up approach still means that senior leaders provide the vision and direction. The difference is that they do not give specific, tactical directives. Then, employees at the lower levels of the organization are expected to apply that direction to their specific area, and use it as the basis for making more specific plans. Those plans are then sent up through the layers of the organization for review and approval. A somewhat extreme example of this approach was taken by Nobuyuki Idei when he was president of Sony. He adopted the motto of "Digital Dream Kids."[12] Western observers all scratched their heads at this, and wondered how something so undefined could serve as an adequate rallying cry to the troops. But the idea was that employees in each part of the organization would think for themselves what this motto meant and how they could reflect it in their activities. Studies show that this type of participatory, bottom-up process is more likely to take a long-term perspective of the organization and enhance capabilities that will contribute to sustainable and long-term performance.[13]

Studies also show that IT—such as ERP and other similar systems—enables bottom-up decision making.[14] The reason is that it provides lower-level employees with information to make good decisions. To achieve sustainable change while becoming a data-driven organization, companies should engage in a participatory leadership approach. And as they become more analytical, they will increasingly become more participatory. Let's now look at the process of change.

10.2 The Four-Step Change Process

There are four essential steps in the change management process: recognize need, unfreeze, change, and refreeze.

10.2.1 Step 1—Recognize Need

The first step in the change process is recognizing the need. This comes from leadership. Opportunities for change come from anywhere in the environment, from both internal and external sources, as well as formal or informal channels. The need for change may be triggered by leaders recognizing that the business environment has changed. For example, there could be a depletion of natural raw materials the company relies upon as inputs, such as coal and oil for energy companies; talent with specific skills may be in short supply, as is the case with analytics capability; there may be new directives from shareholders, such as the need for more data and process transparency. Finally, organizations may need to respond to technological innovations introduced by competitors or suppliers. This is the case with big data analytics.

Organizational leaders may recognize the need to change due to external factors, such as a shift in consumer buying habits. Many companies, for example, have recently discovered the value of data from social media. Telecom companies have found that information from social networks is useful in predicting customer behavior. They discovered that customers who know others who have stopped using a certain telecom are more likely to do so themselves. Using various mechanisms to identify such social networks and tag customers helps track and identify them. These customers are then aggressively targeted for retention programs.

Leaders may identify the need to change from their current data capability and competitive pressures, enabling the creation of entirely new lines of business. Consider that engine-maker Rolls-Royce completely transformed its business over the past decade by analyzing the data from its products, not just building them. Rolls-Royce monitors 3,700 jet engines worldwide.[15] The manufacturer used big data to essentially become a service provider. From its operations center in Britain, the company continuously monitors the performance of its jet engines worldwide and has algorithms that can spot problems before breakdowns occur. The company has used data to help expand its manufacturing business into services—called *servitization*. Here, Rolls-Royce sells the engines but also provides the service to monitor the engines. It then charges customers based on usage time—and repairs or replaces the engines in case of problems. Services now account for around 70 percent of the civil-aircraft engine division's annual revenue.[16] For Rolls-Royce, this was a transformational change.

Consider another example of transformational change—the cement company CEMEX. CEMEX now sells delivery time, not just cement.[17] CEMEX realized that it could increase market share and charge a premium to time-conscious contractors by reducing delivery time on orders. It equipped most of its concrete mixing trucks in Mexico with GPS locators and uses predictive analytics to improve its delivery process ease. This approach allows dispatchers to cut the average response time and increase truck productivity. The company essentially changed its focus from the sale of a commodity to the sale of something customers really care about—time. In short, the unit of business shifted from cubic yards to the delivery window.

Recognizing the need to become a data-driven organization is recognizing the need for transformational change. As big data is implemented, further recognitions for change may develop. Big data enables enterprises of all kinds to create new products and services, enhance existing ones, and invent entirely new business models, as seen with CEMEX. Innovating new business models, products, and services can come from big data analytics.

10.2.2 Step 2—Unfreeze

Unfreezing is preparing the organization for change. Implementing change—such as becoming a data-driven organization—is substantial. We want these changes to be sustained and have long-term impact. We want to avoid resistance from employees and worst of all—sabotage.

Research repeatedly shows that when employees perceive change as a positive opportunity, they increase their support for change and decrease their support for the status quo.[18] However, when they perceive threats, they may increase their commitment to the status quo. Researchers sometimes call this the *threat-rigidity* response.[19] This response can be described as being "frozen" to the current way of doing their work. Although it may be easier to persuade others to make incremental changes, that is often not the case for transformational change. For transformational change to take place, leaders need to plan to "unfreeze" the organization. The primary objectives in the unfreezing step are for leaders to ensure that all employees understand the need for change. This reduces the resistance to change, possibilities of sabotage, and creates a sense of willingness to be a part of the change process.

10.2.2a The Threat—Not a Good Way

During the unfreezing stage of the change process, leaders must convince employees there is a need for change. Change requires employees to implement different behaviors and work patterns. If change is perceived as bad or harmful in some way, employees will likely resist. The goal here is to provide all employees with information that reduces uncertainty. One common influence tactic used to unfreeze organization members is to draw attention to a threat that demands change. This approach can create a sense of discomfort and fear among members as it identifies alarming information or potentially negative consequences of staying with the status quo.

10.2.2b The Opportunity—A Better Way

The second approach is to show the employees that they will be better off if they change. This tactic is to present change as an opportunity that the change offers. This is a positive, feel-good approach that draws attention with a compelling picture of the future that shows not only benefits to the organization, but also personal benefits of the change. An example is to present the change as providing everyone with greater opportunities for growth or advancement or offering the potential for increased financial rewards. Application of big data analytics throughout the organization certainly provides these rewards.

The threat-based approach uses the potential for pain or fear to push people to change. An example is "We are going to lose competitive position, market share, and we will have to create massive layoffs." The opportunity-based approach uses the positive potential to pull people to change. An example would be "Implementing analytics into organizational processes will make us more efficient, make our jobs easier, stocks will go up and we will receive bonuses, plus everyone will acquire more competitive skills." Both approaches to unfreezing are based on creating a sense for the need to change. However, research shows that the threat-based approach may only result in just enough change to avoid a looming painful outcome—just enough for the threat to subside. In this case, the effort is just enough to get by and may not result in employees fully embracing the change. Also, if employees perceive the change itself to be a threat—such as endangering their job, work relationships, compensation, or continued employment—they may quit or become disengaged.[20]

The top-down leadership approach to change involves leaders using their power to influence members to accept their vision and the need for change. This has been shown to not be as sustainable and effective as the bottom-up participatory approach. With the bottom-up process, everyone has become sensitized to the need for the change, creating a readiness to change and a powerful sense of ownership.[21] As management guru Tom Peters says, "When you allow people to share in the process of gathering information, it becomes theirs."[22]

10.2.3 Step 3—Change

Once the organization has gone through unfreezing, it is ready for the actual change. In this stage, the change ideas are put into practice and the change becomes a tangible reality. Change takes the form of interventions, planned activities targeting specific outcomes, such as improving individual, group, or organizational performance and well-being. Implementing big data analytics and moving the organization to become data driven requires interventions across all three organizational areas identified in Chapter 9, "Making It Work": people, processes, and technology. These barriers are actually opportunities for change (see Figure 10.2).

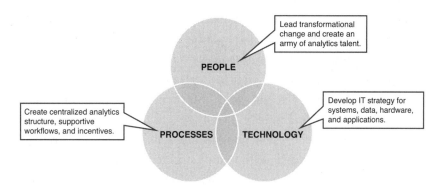

Figure 10.2 The barriers are opportunities. Target areas for intervention are people, processes, and technology

10.2.3a Change Technology

Interventions to change technology are at the crux of becoming an analytics-driven organization. Technological interventions are designed to change the process by which organizational members create knowledge and make decisions. They focus on every aspect of the organization, including workflows, production methods, equipment, and information systems. They are designed to make products or services more efficient and to increase organizational capabilities. Marathon Oil, for example, uses information technology in its purchasing function to reduce inefficiencies in tracking purchases, increase the speed of the bidding process, and save time and money by holding online auctions.[23]

To capture value from big data, organizations will have to deploy a range of new technologies—from storage and computing to analytical software. They will also have to develop capabilities to conduct new techniques, such as new types of analyses as discussed in Chapter 1, "A Game Changer." New business challenges will emerge and—coupled with growing computing power—will fuel the development of new analytical techniques. In addition to improving analytical capabilities, organizations will need to develop mechanisms to visualize, consume, and integrate the large volumes of data across the organization.

As companies move forward with technological implementation, they will have to address issues of legacy systems and siloed IT systems. These often have incompatible standards and formats that prevent the integration of data and prevent the use of more sophisticated analytics. To enable transformative opportunities, companies will increasingly need to integrate information from multiple data sources. Technological intervention here will involve two important decisions. First is the development and rollout of an IT strategic plan. Such a strategy will outline a plan as to whether the organization will be investing in new systems or revamping existing systems and identifying the types of capabilities needed, the time period for implementation, and the extent of third-party involvement. Second, organizations will have to address the issue of data availability and access. They need to have a plan as to whether this data will be purchased, obtained from other supply chain partners, or whether it is something the company already has. Data has become a key competitive asset. Therefore, organizations need to make sure they have access to it and guard what they have closely.

The rate at which these technological interventions take place will depend upon the maturity map of the organization discussed in Chapter 9. However, companies need to proceed with caution rather than jumping in. They should avoid the hype and begin with an assessment of their own needs, identifying gaps in the technology their enterprise has available.

10.2.3b Change Processes

Interventions to change organizational processes are directed at changing processes and systems. Changes to structures include the formal systems of the organization, such as how people are hired and compensated. This may mean addressing issues of organizational hierarchy, chain of command, and how work is carried out. Examples are centralization, standardization, specialization, or departmentalization. For instance, in 2009, the giant computer manufacturer Dell acquired Perot Systems, a service-based information technology company. Suddenly, Perot Systems found itself having to change its processes—from integrated customized service solutions to Dell's focus on standardization.[24]

Process changes may also include changes to the physical environment, such as location of offices and physical access to resources. Interventions to change in the physical setting involve altering the space where people work. This may include types of buildings, the surrounding grounds, how offices or cubicles are configured, and the overall aesthetics of the organization's physical characteristics. The physical setting can have enormous impact on work performance. It affects communication, collaboration, creativity, productivity, and stress levels. For example, Nike created a physical space within the company called the Innovation Kitchen. It is an open space "creative playpen, with every type of tool, material, machine, toy, instrument, software, games and inspiration image at the ready."[25] Similar examples are offered by Google and the SAS Institute, both of which have won awards for their physical space and amenities.[26] According to Google, its aim is "to create the happiest, most productive workplace in the world."[27]

10.2.3c Change People

One of the biggest mistakes companies can make is to upgrade their IT systems—such as adding advanced modules of ERP or CRM systems—but make no changes to the skills of the people who actually use these systems. People are ultimately the ones making decisions. They need to understand how to use analytics. This requires creating an "army" of analytics talent—ranging from highly trained analysts to people who interact with customers but understand the capabilities. This may mean hiring new people and retraining current management and staff. All indicators show that deep analytical talent is in short supply and this is an area companies will need to nurture.

As a result, organizations may need to focus on training the talent they have in place. It may involve sending employees to take university courses or having academic faculty provide in-house training. There are also many novel applications to help integrate analytics into decision making and break down employee resistance. One such approach is being used by the motor carrier Schneider National.[28] The company has developed a simulation game to communicate the importance of analytical thinking in dispatching trucks and trailers. The goal of the game is to minimize costs while maximizing the driver's time on the road. Players make decisions such as whether to accept loads or move empty trucks—all with the help of decision-support tools. The company uses the game to help employees understand the value of analytical decision aids.

Lastly, it is ultimately people who make decisions based on big data analytics. As human beings, we have limited ability to consume and understand data, and big data techniques can greatly assist human decision making. Organizations should use applications to help individuals and teams synthesize information from big data. For instance, more sophisticated visualization techniques, algorithms, and dashboards can enable people to see patterns in large data sets, which can help reveal the most pertinent insights. This tandem of technology and people will get the most out of big data analytics. Recall how this is done with IBM's Watson, providing doctors with alternative treatments with probabilities of outcomes—but leaving the final decision up to the physicians.[29]

10.2.3d Aligning The Building Blocks

The most successful organizational change occurs when all three building blocks—technology, processes, and people—are aligned. When multiple changes are aligned with each other, they have the greatest positive effect. When these changes support each other, they are much more likely to result in lasting change. Consider that changes made to analytics applications, IT implementation, and organizational processes have a direct effect on people. If people are not prepared to utilize the technologies implemented, if they cannot access data, if analytics is not integrated into workflow, if they do not have the skills to use the technology, and if there are not incentives—then lasting change will not take place. All three must be coordinated through an overarching strategy and leadership.

One way to begin to facilitate an understanding of supply chain analytics is through simpler applications with narrow functionality. These are increasingly referred to as *analytical apps*. They are easy-to-understand tools that are similar to the applications found on smartphones. They support a single decision and often are industry-specific. Several business intelligence and analytics software vendors are introducing them, and they promise to make the use of analytics much simpler and available to users who do not have extensive analytical or technological skills. Analytical apps that have already been developed for supply chain functions include tools for supplier evaluation, inventory performance analysis, transportation analytics, and transportation contract compliance. There undoubtedly will be many others over the next several years.

When analytics applications are easily accessible, they are used. Consider UPS. All their drivers use On-Road Integrated Optimization and Navigation (ORION).[30] It is a simple-to-use navigation tool that helps drivers find the most efficient path through their delivery area. It provides drivers with the most efficient route based on data, analytics, algorithms, and modeling. The drivers, however, don't need to understand the modeling or algorithms behind the device. They just have a simple, easy-to-use device they trust.

10.2.4 Step 4—Refreeze

Once changes have been implemented, they need to be reinforced in order to become institutionalized. Otherwise, people will easily slip back into old habits. This is called "refreezing." It refers to ensuring that the change becomes embedded in everyday actions and habits. Achieving this goal requires creating structures and systems that reinforce the change and dismantling those that undermine it.

It is in the refreezing stage that leaders need to make adjustments to the design of the organization—the people, processes, and technology. Major change has already occurred and everyone is already committed to the change. What happens here are incremental changes that refine and make the system even better. This may include structural changes to the way work is organized, such as how IT is integrated into certain work processes or the reporting structure. It may require training and making changes to the incentive systems to solidify these behaviors. This may also require promoting the new practices to solidify them. Basically, this final step of refreezing is merely reinforcing and solidifying changes structurally and systematically to ensure that the new ways of doing things are repeated and rewarded.

A more contemporary approach to ensuring that positive changes are implemented throughout the organization is referred to as "re-slushing."[31] The idea here is that it is more sustainable to re-slush rather than refreeze the organization, in order to facilitate ongoing experimentation and changes. Here, greater emphasis is placed on changes that facilitate learning and flexibility. Also, re-slushing suggests that change is an ongoing process rather than a one-time event. As technology evolves, flexible systems allow ongoing change and continuous improvement to take place.

10.3 Leadership

Only senior leadership can spearhead the transformational change required to build and apply analytical capabilities across their organizations. This transformational change *will not* happen if the responsibility is abdicated to middle managers, professional analysts, or external consultants.

It is leadership at driven companies that has put analytics at center stage. This is underscored by the words of Gavin Gallagher, chief information officer of H-E-B Grocery Stores, who was quoted as saying, "Analytics are king at H-E-B. They are where I plan to spend most of my money. There is much more competitive advantage with analytics than in installing additional transaction systems."[32] Similarly, Timothy Carroll, vice president of supply chain operations at IBM, has said, "We are constantly playing out scenarios through business analytics to determine if we have a way of quickly recovering from a situation."[33] The commitment to analytics is so strong at Amazon that CEO Jeff Bezos had reportedly fired a group of Web designers who relied on fad and fashion in Web design rather than metrics and analysis.[34]

It is top leadership that needs to provide this level of commitment, involvement, and vision to analytics in order to move the organization through the needed transformation. It is during the periods of change—as described in the change process—that the role of leader moves from mere words to actions. Leaders need to become *change agents*—individuals who act as a catalyst and take leadership and responsibility for managing the change process. Change agents make things happen. They create the vision. They provide the resources, and create the structure for changes to happen to technology, process, and people. They also need to exhibit appropriate skills, clarify expectations, and ensure that training is provided. Recall that Hiroshi Mikitani, CEO of Rakuten, realized he had to provide training and education if he was to transform his organization into an English-speaking entity. Leaders also need to communicate intrinsic and extrinsic benefits to employees—especially during the unfreezing process.

Change agents often are most successful when they work with *idea champions*, people who actively and enthusiastically support new ideas. Together, change agents and idea champions promote productive change within the organization by building support, overcoming resistance, and ensuring that innovations are implemented. For transformational change to take place, there needs to be commitment from all organizational members. Commitment is important because it impacts the behaviors and performance of individual members. Only senior leadership can make this happen.

10.3.1 What Should Leaders Do?

Research has shown that employees of an organization will be more committed to change if they have confidence in the competencies of their leaders. There are three ways this confidence can develop. First, leaders need to lead by example and exhibit appropriate skills (see Table 10.1). A leader who wants to lead an analytics-driven organization needs to exhibit at least a rudimentary level of analytical literacy. Leaders certainly don't have to be experts, but just demonstrate that they value fact-based decisions. Second, leaders need to make sure that expectations are clear and that everyone has the right training. Last, leaders need to ensure there are both extrinsic and intrinsic rewards.

Table 10.1 Leadership Behaviors of a Change Agent

LEADERS SHOULD		
Exhibit appropriate skills	Clarify expectations	Communicate intrinsic benefits
Lead by example	Provide training	Link to extrinsic reward
Demonstrate credibility	Plan for success	Select or promote positive attitudes

Organizations that seek a competitive advantage from big data analytics must take an enterprise-wide perspective in their applications. Leaders need to ensure that analytics takes a cross-functional, cross-product, cross-customer, cross-supply chain approach. This is the only way analytics will make a difference in the organization. Many analytics applications are currently embedded in a set of organizational silos. Leaders need to ensure that silos are broken down and that analytics applications connect detailed information on product availability with the models that predict demand, so the interaction between demand and supply can be better understood. However, this is not enough for a competitive advantage. Business leaders need to know how to take advantage of the big data capability they are creating. Leaders have to have some understanding and appreciation of analytical capability and techniques. Big data analytics has demonstrated its potential. It is now up to leaders to lead their organizations through the transformation.

Endnotes

Chapter 1

1. Todd Leopold, "The Death and Life of a Great American Bookstore," CNN, September 12, 2011; available at http://www.cnn.com/2011/US/09/12/first.borders.bookstore.closing/

2. The term "datafication" is introduced by Victor Mayer-Schönberger and Kenneth Cukier, *Big Data: A Revolution That Will Transform How We Live, Work, and Think* (New York: Houghton Mifflin Harcourt, 2013); it means to put data in quantified form so it can be tabulated and analyzed (p. 78).

3. Thomas H. Davenport, "Competing on Analytics," *Harvard Business Review* (January 2006): 1–9.

4. Thomas H. Davenport and Jeanne G. Harris, *Competing on Analytics: The New Science of Winning* (Boston, MA: Harvard Business Publishing Corporation, 2007).

5. There are many definitions of big data; one of the best is by J. Maniyaka et al., "Big Data: The Next Frontier for Innovation, Competition, and Productivity," McKinsey Global Institute White Paper, May 2011.

6. "A Different Game," *The Economist,* February 27, 2010, pp. 6–8.

7. Thomas H. Davenport, "Realizing the Potential of Retail Analytics: Plenty of Food for Those with the Appetite," Working Knowledge Research Report, Babson Executive Education White Paper, 2009, pp. 1–42.

8. Thomas H. Davenport, "Competing on Analytics," *Harvard Business Review* (January 2006): 1–9.

9. Thomas H. Davenport, "Realizing the Potential of Retail Analytics: Plenty of Food for Those with the Appetite," Working Knowledge Research Report, Babson Executive Education White Paper, 2009, pp. 1–42.

10. Andrew McAfee and Erik Brynjolfsson, "Big Data: The Management Revolution," *Harvard Business Review* (October 2012).

11. Moore's law states that the number of transistors on a chip roughly doubles every two years.

12. J. Maniyaka et al., "Big Data: The Next Frontier for Innovation, Competition, and Productivity," McKinsey Global Institute White Paper, May 2011.

13. Abhishek Mehta, "Big Data: Powering the Next Industrial Revolution," Tableau Software White Paper, 2011 (http://www.tableausoftware.com/learn/whitepapers/big-data-revolution).

14. Steve Lohr, "New Ways to Exploit Raw Data May Bring Surge of Innovation, a Study Says," *New York Times*, May 13, 2011.

15. This is one way to think about big data. Others are adding a fourth term, such as *veracity* to incorporate accuracy. One source is by Edd Dumbill, "Volume, Velocity, Variety: What You Need to Know About Big Data," *forbes.com*, January 19, 2012.

16. There are many statistics documenting the amount of data. Some good sources include:

    J. Maniyaka et al., "Big Data: The Next Frontier for Innovation, Competition, and Productivity," McKinsey Global Institute White Paper, May 2011.

    Victor Mayer-Schönberger and Kenneth Cukier, *Big Data, A Revolution That Will Transform How We Live, Work, and Think* (New York: Houghton Mifflin Harcourt, 2013).

17. Ibid.

18. J. Maniyaka et al., "Big Data: The Next Frontier for Innovation, Competition, and Productivity," McKinsey Global Institute White Paper, May 2011.

19. A term coined by Kevin Ashton in "That 'Internet of Things' Thing, in the Real World Things Matter More Than Ideas," *RFID Journal* (22 June 2009).

20. www.softwareconsortium.com/the-era-of-big-data.html

21. J. Maniyaka et al., "Big Data: The Next Frontier for Innovation, Competition, and Productivity," McKinsey Global Institute White Paper, May 2011.

22. Victor Mayer-Schönberger and Kenneth Cukier, *Big Data, A Revolution That Will Transform How We Live, Work, and Think* (New York: Houghton Mifflin Harcourt, 2013, p. 59).

23. There are many books on analytics. One recent source is by James Evans, *Business Analytics: Methods, Models, and Decisions* (Upper Saddle River, NJ: Pearson, 2013).

24. Jessi Hempel, "IBM's Massive Bet on Watson," *Fortune*, October 7, 2013, pp. 81–88.

25. Victor Mayer-Schönberger and Kenneth Cukier, *Big Data, A Revolution That Will Transform How We Live, Work, and Think* (New York: Houghton Mifflin Harcourt, 2013, p. 59).

26. Jeremy Ginsburg et al., "Detecting Influenza Epidemics Using Search Engine Query Data," *Nature 457* (2009): 1012–1014, http://www.nature.com/nature/journal/v457/n7232/full/nature07634.html.

27. The term was coined in the book by Nassim Nicholas Taleb, *The Black Swan* (U.K.: Random House, 2007).

28. Kashmir Hill, "How Target Figured Out a Teen Girl Was Pregnant Before Her Father Did," *Forbes*, February, 12, 2012.

29. Thomas H. Davenport, "Realizing the Potential of Retail Analytics: Plenty of Food for Those with the Appetite," Working Knowledge Research Report, Babson Executive Education White Paper, 2009, pp. 1–42.

30. Ibid.

31. Peter Katel, "Bordering on Chaos," *Wired*, March 14, 2012, pp. 1–6.

32. James Cooke, "Running Inventory Like a Deere," *CSCMP Supply Chain Management Quarterly* (February 14, 2007).

33. "Inside P&Gs Digital Revolution," *McKinsey Quarterly* (November 2011).

34. Kwame Opam, "Amazon Plans to Ship Your Package Before You Even Buy Them," *The Verge*, January 18, 2014.

35. "2012 Shopper Engagement Study: Media Topline Report," Point of Purchase Advertising International Report (Chicago, IL, 2012); "Shopper Marketing Best Practices: A Collaborative Model for Retailers and Manufacturers," Retail Commission on Shopper Marketing: The Partnering Group Report (Cincinnati, OH, 2010); V. Shankar, J. J. Inman, M. Mantrala, E. Kelley, and R. Rizley, 2011. "Innovations in Shopper Marketing: Current Insights and Future Research Issues," *Journal of Retailing 87S, 1* (2011):S29–S42.

36. J. Maniyaka et al., "Big Data: The Next Frontier for Innovation, Competition, and Productivity," McKinsey Global Institute White Paper, May 2011.

37. Victor Mayer-Schönberger and Kenneth Cukier, *Big Data, A Revolution That Will Transform How We Live, Work, and Think* (New York: Houghton Mifflin Harcourt, 2013, pp. 135–136).

38. Mathew Kulp, "Material Handling Equipment for Multichannel Success," *CSCMP Supply Chain Quarterly*, 4 (2013): 32–37.

39. J. Maniyaka et al., "Big Data: The Next Frontier for Innovation, Competition, and Productivity," McKinsey Global Institute White Paper, May 2011.

40. "A Different Game," *The Economist,* February 27, 2010, pp. 6–8.

Chapter 2

1. "Data, data everywhere," *The Economist*, February 25th, 2010.

2. A great deal has been written about Walmart's unifying platform Retail Link. Some good sources are:

   Victor Mayer-Schönberger and Kenneth Cukier, *Big Data, A Revolution That Will Transform How We Live, Work, and Think* (New York: Houghton Mifflin Harcourt, 2013).

   "A Different Game," *The Economist*, February 25, 2010. Also available at http://www.economist.com/node/15557465.

   http://www.pbs.org/wgbh/pages/frontline/shows/walmart/secrets/pricing.html.

3. Stephanie Clifford, "Using Data to Stage-Manage Paths to the Prescription Counter," *Big Data*, June 19, 2013. Also available at http://bits.blogs.nytimes.com/2013/06/19/using-data-to-stage-manage-paths-to-the-prescription-counter/?_php=true&_type=b.

4. Ibid.
   http://cvssuppliers.com/

5. Thomas H. Davenport, "Realizing the Potential of Retail Analytics: Plenty of Food for Those with the Appetite," Working Knowledge Research Report, Babson Executive Education White Paper, 2009, pp. 1–42.

6. A good differentiation among software vendors is offered by Thomas H. Davenport and Jinho Kim, *Keeping Up with the Quants* (Boston, MA: Harvard Business Publishing Corporation, 2013, p. 76).

7. Ibid.

8. Mary Siegried, "Find the Big Picture in Big Data," *Inside Supply Management* (January-February 2014): 19–23. Also available at www.ism.ws.

9. M. R. Leenders, P. F. Johnson, A. E. Flynn, and H. E. Fearon, *Purchasing and Supply Management* (New York, NY: McGraw-Hill/Irwin, 2010, p. 6).

10. Brad Stone, "The Secrets of Bezos: How Amazon Became the Everything Store," *Bloomberg Businessweek*, October 10, 2013, http://www.businessweek.com/articles/2013-10-10-/jeff-bezos-and-the-age-amazon-excerpts-from-the-everything-store-by-brad-stone.

11. Mary Siegried, "Find the Big Picture in Big Data," *Inside Supply Management* (January-February 2014): 19–23. Also available at www.ism.ws.

12. David Kiron, Renee Boucher Ferguson, and Pamela Kirk Prentice, "From Value to Vision: Reimagining the Possible with Data Analytics," *MIT Sloan Management Review Research Report* (Spring 2013).

13. Thomas H. Davenport and Jeanne G. Harris, *Competing on Analytics: The New Science of Winning* (Boston, MA: Harvard Business Publishing Corporation, 2007).

14. "Coke Has a Secret Formula for Orange Juice, Too," *Bloomberg Businessweek*, February 4–10, 2013, pp. 19–21.

15. Adriana Lee, "How Luxury Retailers Are Spying on Shoppers with Surveillance Mannequins," *TechnoBuffalo*, November 24, 2012, http://www.technobuffalo.com/2012/11/24/how-luxury-retailers-are-spying-on-shoppers-with-surveillance-mannequins/.

16. Thomas H. Davenport, "Realizing the Potential of Retail Analytics: Plenty of Food for Those with the Appetite," *Working Knowledge Research Report*, Babson Executive Education White Paper, 2009, pp. 1–42.

17. Joseph C. Andraski, "The Case for Item-level RFID," *CSCMP's Supply Chain Quarterly*, 4 (2013): 46–52.

18. Thomas H. Davenport, "Realizing the Potential of Retail Analytics: Plenty of Food for Those with the Appetite," *Working Knowledge Research Report*, Babson Executive Education, 2009, pp. 1–42.

19. Ibid.

Chapter 3

1. Thomas H. Davenport, "Realizing the Potential of Retail Analytics: Plenty of Food for Those with the Appetite," *Working Knowledge Research Report*, Babson Executive Education, 2009, pp. 1–42.

2. Sanders, Nada, Big Data, *SupplyChainBrain*, January 2013.

3. T. S. Baines, H. W. Lightfoot, O. Benedettini, and J. M. Kay, "The Servitization of Manufacturing: A Review of Literature and Reflection on Future Challenges," *Journal of Manufacturing Technology Management* 20, no. 5 (2009): 547–567.

   T. S. Baines, H. Lightfoot, J. Peppard, M. Johnson, A. Tiwari, E. Shehab, and M. Swink, "Towards an Operations Strategy for Product-Centric Servitization," *International Journal of Operations & Production Management* 29, no. 5 (2009): 494–519.

4. J. Maniyaka et al., "Big Data: The Next Frontier for Innovation, Competition, and Productivity," McKinsey Global Institute White Paper, May 2011.

   David Kiron, Renee Boucher Ferguson, and Pamela Kirk Prentice, "From Value to Vision: Reimagining the Possible with Data Analytics," *MIT Sloan Management Review Research Report*, Spring 2013.

5. Thomas H. Davenport and Jinho Kim, *Keeping Up with the Quants* (Boston, MA: Harvard Business Publishing Corporation, 2013).

J. Maniyaka et al., "Big Data: The Next Frontier for Innovation, Competition, and Productivity," McKinsey Global Institute White Paper, May 2011.

6. Ibid.

7. Author interviews.

8. Jeremy Ginsburg et al., "Detecting Influenza Epidemics Using Search Engine Query Data," *Nature* 457 (2009): 1012–1014, http://www.nature.com/nature/journal/v457/n7232/full/nature07634.html.

9. See Mike Ledyard, "Is Your Metrics Program Measuring Up?" *CLM Explores* Vol. 1 (Spring/Summer 2004): 7.

Chapter 4

1. John Furrier, "Bit Data Is Big Market & Big Business—$50 Billion Market by 2017," *Forbes*, February 17, 2012.

2. Kashmir Hill, "How Target Figured Out a Teen Girl Was Pregnant Before Her Father Did," *Forbes*, February, 12, 2012.

3. Charles Duhigg, "How Companies Learn Your Secrets," *New York Times*, February 16, 2012, http://www.newyorktimes.com.

4. Patrick Vlaskovits, Henry Ford, Innovation, and That "Faster Horse" Quote, Harvard Business Review Blog, August 29, 2011, available at http://blogs.hbr.org/2011/08/henry-ford-never-said-the-fast/

5. Excellent examples are offered by Thomas H. Davenport, "Realizing the Potential of Retail Analytics: Plenty of Food for Those with the Appetite," *Working Knowledge Research Report*, Babson Executive Education White Paper, 2009, pp. 1–42.

6. "Netsertive Powers Home Technology Specialists of America (HTSA) Holiday Ad Campaign," http://www.netsertive.com, November 29, 2010.

7. "From Art to Science: How Retail Analytics Can Assist with Price Optimization," http://www.saleswarp.com/product-management, March 18, 2013.

8. Thomas H. Davenport, "Realizing the Potential of Retail Analytics: Plenty of Food for Those with the Appetite," *Working Knowledge Research Report*, Babson Executive Education White Paper, 2009, pp. 1–42.

9. Paul Demery, "How Analytics Support Site Design at Overstock.com," *Internet Retailer*, February 22, 2007.

10. Thomas H. Davenport, "Realizing the Potential of Retail Analytics: Plenty of Food for Those with the Appetite," *Working Knowledge Research Report*, Babson Executive Education White Paper, 2009, pp. 1–42.

11. Ibid.

12. Jim Tierney, "Instagram Contest Kicks off 30th Anniversary of Neiman Marcus Loyalty Program," *Loyalty 360.org*, January 27, 2014.

13. Taddy Hall, "Customers Don't Care About Loyalty Programs as Much as Brands Think They Do," http://www.businessinsider.com, April 4, 2013.

14. This has been published in numerous analyst reports, such as "Building with Big Data: The Data Revolution Is Changing the Landscape of Business," *The Economist*, May 26, 2011, http://www.economist.com/node/18741392/.

15. Netflix, Netflix Prize, available at http://www.netflixprize.com

16. Shel Israel, "How Wal-Mart and Heineken Will Use Shopperception to Put Your In-Store Experience in Context," *Forbes*, January 27, 2013, http://www.forbes.com.

17. Nigel Hollis, "Starbucks' New Logo: A Risky Move," HBR Blog Network, January 7, 2011, http://www.blogs.hbr.org/2011/01/starbucks-new-logo-apple-or-le/.

18. Thomas H. Davenport, "Realizing the Potential of Retail Analytics: Plenty of Food for Those with the Appetite," *Working Knowledge Research Report*, Babson Executive Education White Paper, 2009, p. 17.

19. Retail Analytics: Case Illustration, AAUM Research & Analytics White Paper, December 2013, available at http://www.aaumanalytics.com/Presentations/Retail.pdf

20. Ibid.

21. Thomas H. Davenport, "Realizing the Potential of Retail Analytics Plenty of Food for Those with the Appetite," *Working Knowledge Research Report*, Babson Executive Education, 2009, pp. 1–42.

22. Mobile Industry Statistics, Digby, available at http://digby.com/mobile-statistics/

23. Solutions, Placecast, available at http://placecast.net/shopalerts/operators.html

24. About Street Bump, Street Bump, http://streetbump.org/about/

25. CitySense, Sense Networks, available at https://www.sensenetworks.com/products/macrosense-technology-platform/citysense/

26. This quote is attributed to John Wanamaker, U.S. department store merchant (1838–1922).

27. Thomas H. Davenport, "Realizing the Potential of Retail Analytics: Plenty of Food for Those with the Appetite," *Working Knowledge Research Report*, Babson Executive Education White Paper, 2009, pp. 1–42.

28. Ravi Kalakota, "Big Data, Analytics and KPIs in E-commerce and Retail Industry," PracticalAnalytics, August 3, 2011, http://practicalanalytics.worldpress.com/2011/08/03/big-data-analytics-and-kpis-in-e-commerce-and-retail-industry/.

29. Tom French, Laura LaBerge, and Paul Magill, "We're All Marketers Now," Insights & Publications, McKinsey & Company, July 2011, http://www.mckinsey.com/insights/marketing_sales/were_all_marketers_now.

Chapter 5

1. Hessman, Travis, "Putting Big Data to Work," *Industry Week*, April 2013, pp. 14–18, http://www.industryweek.com.

2. This is a number with 10^{24}. Essentially, it is a huge amount of data.

3. A large amount of literature exists on the emergent trend of servitization. Some research articles include the following: T. S. Baines, H. W. Lightfoot, O. Benedettini, and J. M. Kay, "The Servitization of Manufacturing: A Review of Literature and Reflection on Future Challenges," *Journal of Manufacturing Technology Management 20* no. 5 (2009): 547–567; T. S. Baines, H. Lightfoot, J. Peppard, M. Johnson, A. Tiwari, E. Shehab, and M. Swink, "Towards an Operations Strategy for Product-Centric Servitization," *International Journal of Operations & Production Management 29* no. 5 (2009): 494–519; J. Spohrer and P. P. Maglio, "The Emergence of Service Science: Toward Systematic Service Innovations to Accelerate Co-creation of Value," *Production and Operations Management 17* no. 3 (2008): 238–246.

4. M. Hopkins, "How to Innovate When Platforms Won't Stop Moving," *Sloan Management Review 52* no. 4 (2011): 54–60.

5. M. W. Johnson, C. C. Christensen, and H. Kagermann, "Reinventing Your Business Model," *Harvard Business Review 86* (2008): 50–59.

6. Travis Hessman, "Putting Big Data to Work," *Industry Week*, April 2013, pp. 14–18, http://www.industryweek.com.

7. Ibid.

8. J. Maniyaka et al., "Big Data: The Next Frontier for Innovation, Competition, and Productivity," McKinsey Global Institute White Paper, May 2011.

Henry Blodget & Alex Cocotas, The State of the Internet, Business Insider, October 1, 2012, available at http://www.businessinsider.com/state-of-internet-slides-2012-10?op=1

A Different Game, The Economist Soecial Report, February 25, 2010, available at http://www.economist.com/node/15557465

9. Featured Industry Guides, Shopper Marketing, available at http://www.shoppermarketingmag.com/home

10. J. Maniyaka et al., "Big Data: The Next Frontier for Innovation, Competition, and Productivity," McKinsey Global Institute White Paper, May 2011.

11. Ibid.

12. Ibid.

13. Ibid.

14. Ibid.

15. "Online Extra: The Secret of BMW's Success," *Bloomberg Businessweek*, October 15, 2006.

16. Nada R. Sanders, *Supply Chain Management A Global Perspective* (New York: John Wiley & Sons, 2012).

17. R. Dan Reid and Nada R. Sanders, *Operations Management*, 5th edition (New York: John Wiley & Sons, 2012).

18. Thomas H. Davenport and Jeanne G. Harris, *Competing on Analytics: The New Science of Winning* (Boston, MA: Harvard Business School Publishing Corporation, 2007).

    Cindy Etsell, "Analyze This: Enhancing Operational Excellence Through Customer Data," *Digital Manufacturing*, February 15, 2012, http://www.manufacturingdigital.com/people_skills/analyse-this-enhancing-operational-excellence-through-customer-data.

19. Travis Hessman, "Putting Big Data to Work," *Industry Week*, April 2013, pp. 14–18, http://www.industryweek.com.

20. Thomas H. Davenport, "Realizing the Potential of Retail Analytics: Plenty of Food for Those with the Appetite," *Working Knowledge Research Report*, Babson Executive Education White Paper, 2009, pp. 1–42.

21. "Best Practices in Customer Service and Store Performance Management," *Aberdeen Group*, September, 2005.

22. Paul Brody, "Get Ready for the Software-Defined Supply Chain," *CSCMP's Supply Chain Quarterly 4* (2013): 27–30.

    "The New Software-Defined Supply Chain—Preparing for the Disruptive Transformation of Electronics Design and Manufacturing," *IBM Institute for Business Value*, 2013, http://www-935.ibm.com/services/us/gbs/thoughtleadership/software-defined-supply-chain/. Also available on the IBM IBV app at https://itunes.apple.com/us/app/ibm-ibv/id490117648?mt=8.

23. Paul Brody, "Get Ready for the Software-Defined Supply Chain," *CSCMP's Supply Chain Quarterly 4* (2013): 27–30.

24. Victor Mayer-Schönberger and Kenneth Cukier, *Big Data, A Revolution That Will Transform How We Live, Work, and Think* (New York: Houghton Mifflin Harcourt, 2013, p. 40).

25. Adam Elkus, "McKinsey on Big Data Part 2: Production, Supply, and Logistics," December 16, 2011, http://www.ctovision.com/2011/12/mckinsey-on-big-data-part-2-production-supply-and-logistics/.

26. Ibid

27. Ibid.

Chapter 6

1. Adapted from "Harry Potter and the Logistical Nightmare," *Business Week*, August 6, 2007, 9.

2. George A. Gecowets, "Physical Distribution Management," *Defense Transportation Journal* 35, no. 4 (August 1979): 5.

3. Supply Chain Visions, *Supply Chain Management Process Standards* (Oak Brook, IL: Council of Supply Chain Management Professionals, 2004, p. 25).

4. Peter F. Drucker, "The Economy's Dark Continent," *Fortune* 65, no. 4 (April 1962): 103.

5. See Mark Schlack, "Blending the New with the Old," in *Technology Priorities for 2013*, a TechTarget e-publication (April 2013), pp. 2–4.

6. See Brenda Cole, "Manufacturing Mobility Rises While MDM Adoption Lags Behind," in *Technology Priorities for 2013*, a TechTarget e-publication (April 2013), pp. 13–16.

7. Thomas H. Davenport, "Realizing the Potential of Retail Analytics: Plenty of Food for Those with the Appetite," *Working Knowledge Research Report*, Babson Executive Education White Paper, 2009, pp. 1–42.

8. Ibid.

9. Claire Swedberg, "Daisy Brand Benefits from RFID Analytics," *RFID Journal*, January 18, 2008, http://www.rfidjournal.com/articles/views?3860.

10. Ibid.

11. Victor Mayer-Schönberger and Kenneth Cukier, *Big Data, A Revolution That Will Transform How We Live, Work, and Think* (Boston, MA: Houghton Mifflin Harcourt, 2013).

12. Bernard J. La Londe and Paul H. Zinszer, *Customer Service: Meaning and Measurement* (Chicago, IL: National Council of Physical Distribution Management, 1976).

13. Victor Mayer-Schönberger and Kenneth Cukier, *Big Data, A Revolution That Will Transform How We Live, Work, and Think* (Boston, MA: Houghton Mifflin Harcourt, 2013).

14. Joseph C. Andraski, "The Case for Item-Level RFID," *CSCMP's Supply Chain Quarterly* 4 (2013): 46–52.

15. Mathew Kulp, "Material Handling Equipment for Multichannel Success," *CSCMP's Supply Chain Quarterly* 4 (2013): 32–37.

16. Marc Wulfraat, "Automated Case Picking for Grocery Distribution," *Supply Chain Digest*, June 27, 2013, http://www.scdigest.com/experts/Wulfraat_13-06-27.php?cid=7191.

17. Ibid.

18. Picking, Kiva Systems, available at http://www.kivasystems.com/solutions/picking/

19. Thomas H. Davenport, "Realizing the Potential of Retail Analytics: Plenty of Food for Those with the Appetite," *Working Knowledge Research Report*, Babson Executive Education White Paper, 2009, pp. 1–42.

20. Ibid.

21. Joseph C. Andraski, "The Case for Item-Level RFID," *CSCMP's Supply Chain Quarterly* 4 (2013): 46–52.

22. Dale S. Rogers, Ron Lembke, and John Benardino, "Taking Control of Reverse Logistics," *Logistics Management* 52, no. 5 (May 2013): 54.

23. Inmar Operates Intelligent Commercial Networks, Inmar, available at http://www.inmar.com/

24. Leslie Guevarra, "Unilever Tops List of Sustainability Leaders," GreenBiz.com, April 10, 2011, http://www.greenbiz.com/print/42337.

25. Ibid.

26. Manufacturing Energy Consumption Survey (MECS), U.S. Energy Information Administration, available at http://www.eia.gov/consumption/manufacturing/index.cfm

27. Mathew Naitove, "Collaborative Robot Works Safely, Comfortably, Alongside Humans," *Plastics Technology*, November 2013.

Chapter 7

1. James Cook, "From Many, One: IBM's Unified Supply Chain," *CSCMP Supply Chain Quarterly* 4 (2012).

2. Ibid.

3. M. R. Leenders, P. F. Johnson, A. E. Flynn, and H.E. Fearon, *Purchasing and Supply Management* (New York, NY: McGraw-Hill/Irwin, 2010, p. 6).

4. Jason Busch, The Meaning of Big Data for Procurement and Supply Chain: A Fundamental Information Shift, Spend Matters,

available at http://spendmatters.com/2012/05/18/the-meaning-of-big-data-for-procurement-and-supply-chain-a-fundamental-information-shift/#sthash.WgV3QTJG.dpuf

5. James Cook, "From Many, One: IBM's Unified Supply Chain," *CSCMP Supply Chain Quarterly* 4 (2012).

6. J. Maniyaka et al., "Big Data: The Next Frontier for Innovation, Competition, and Productivity," McKinsey Global Institute White Paper, May 2011.

   Henry Blodget & Alex Cocotas, The State of the Internet, Business Insider, available at http://www.businessinsider.com/state-of-internet-slides-2012-10?op=1

7. A mash-up is a Web page, or Web application, that uses content from more than one source to create a single new service displayed in a single graphical interface. For example, you could combine the addresses and photographs of your library branches with a Google map to create a map mash-up.

8. Thomas H. Davenport, "Realizing the Potential of Retail Analytics: Plenty of Food for Those with the Appetite," *Working Knowledge Research Report*, Babson Executive Education White Paper, 2009, pp. 1–42.

9. James Cook, "From Many, One: IBM's Unified Supply Chain," *CSCMP Supply Chain Quarterly* 4 (2012).

10. J. Maniyaka et al., "Big Data: The Next Frontier for Innovation, Competition, and Productivity," McKinsey Global Institute White Paper, May 2011.

11. Ibid.

12. H. L. Lee, "Aligning Supply Chain Strategies with Product Uncertainty," *California Management Review* 4493 (Spring 2002): 105–119.

    Marshall Fisher, "What Is the Right Supply Chain for Your Product?" *Harvard Business Review* (March–April 1997): 83–93.

13. Kwame Opam, "Amazon Plans to Ship Your Packages Before You Even Buy Them," *The Verge*, January 18, 2014.

    Thomas H. Davenport, "Realizing the Potential of Retail Analytics: Plenty of Food for Those with the Appetite," *Working Knowledge Research Report*, Babson Executive Education White Paper, 2009, pp. 1–42.

14. James Cook, "From Many, One: IBM's Unified Supply Chain," *CSCMP Supply Chain Quarterly* 4 (2012).

15. Ibid.

16. Thomas H. Davenport, "Realizing the Potential of Retail Analytics: Plenty of Food for Those with the Appetite," *Working Knowledge Research Report*, Babson Executive Education White Paper, 2009, pp. 1–42.

17. N. R. Sanders and A. Locke, "Making Sense of Outsourcing," *Supply Chain Management Review* 9, no. 2 (2005): 38–45.

18. Thomas H. Davenport and Jerry O'Dwyer, "Tap into the Power of Analytics," *CSCMP Quarterly* 4 (2011).

19. David Simchi-Levi, *Operations Rules: Delivery Customer Value through Flexible Operations* (Cambridge, MA: Massachusetts Institute of Technology, 2010).

20. Ibid.

21. U.S. Department of Labor (2013), http://www.dol.gov/

22. Mary Siegried, "Find the Big Picture in Big Data," *Inside Supply Management* (January–February 2014): 19–23, http://www.ism.ws.

23. Deborah Catalano Ruriani, "Improving Your Security Program," *Inbound Logistics* 27, no. 2 (February 2007): 8–9.

24. "12 Million Containers: Can We Scan Them All?" *DC Velocity* 5, no. 3 (March 2007): 11–12.

25. Larry Dignan, "Fedex Launches SenseAware: Collaboration Meets GPS Meets Sensory Data," *Smartplanet*, November 16, 2009.

26. Coral Davenport, "Industry Awakens to Threat of Climate Change," *New York Times*, January 23, 2014.

27. Ibid.

28. Ibid.

29. "Supply Chain Disruptions: From the Warehouse to Wall Street," *Channels* 10, no. 1 (2005): 2.

Chapter 8

1. Thomas H. Davenport and Jeanne G. Harris, *Competing on Analytics: The New Science of Winning* (Cambridge, MA: Harvard Business School Publishing Corporation, 2007).

2. Numerous studies support these characteristics.

   Andrew Mcafee and Erik Brynjolfsson, "Big Data: The Management Revolution," *Harvard Business Review* (October 2012).

   Thomas H. Davenport, "Competing on Analytics," *Harvard Business Review* (January 2006): 1–9.

   J. Maniyaka et al., "Big Data: The Next Frontier for Innovation, Competition, and Productivity," McKinsey Global Institute White Paper, May 2011.

   Sanders, Nada, "Big Data," *SupplyChainBrain*, January 2013.

3. Victor Mayer-Schönberger and Kenneth Cukier, *Big Data, A Revolution That Will Transform How We Live, Work, and Think* (Boston, MA: Houghton Mifflin Harcourt, 2013).

4. J. Maniyaka et al., "Big Data: The Next Frontier for Innovation, Competition, and Productivity," McKinsey Global Institute White Paper, May 2011.

Thomas H. Davenport, "Competing on Analytics," *Harvard Business Review* (January 2006): 1–9.

5. Ibid.

6. J. Maniyaka et al., "Big Data: The Next Frontier for Innovation, Competition, and Productivity," McKinsey Global Institute White Paper, May 2011.

7. Thomas H. Davenport and Jeanne G. Harris, *Competing on Analytics: The New Science of Winning* (Cambridge, MA: Harvard Business School Publishing Corporation, 2007).

8. Thomas H. Davenport, "Competing on Analytics," *Harvard Business Review* (January 2006): 1–9.

9. M. E. Porter, "From Competitive Advantage to Corporate Strategy," *Harvard Business Review* (May/June 1987): 43–59; M. E. Porter, "What Is Strategy," Harvard Business Review (Nov/Dec 1996); M. E. Porter, On Competition (Boston: Harvard Business School, 1998); R. H. Hayes and S. C. Wheelwright, "Link Manufacturing Process and Product Life Cycles," *Harvard Business Review* (January-February 1979): 133–140.

10. "FordDirect Debuts DealerConnection Elite Service for Dealers at the National Automobile Dealers Association (NADA) Convention," *PR Newswire*, February 4, 2012.

11. Thomas H. Davenport, "Realizing the Potential of Retail Analytics: Plenty of Food for Those with the Appetite," *Working Knowledge Research Report*, Babson Executive Education White Paper, 2009, pp. 1–42.

12. Trendsetting Clothing Retailer Maximizes Returns Across Brands and Channels Using Adobe Marketing Cloud, American Eagle Outfitters, available at http://www.images.adobe.com/.pdfs/americaneagle-case-study.pdf

13. R. Dan Reid and Nada R. Sanders, *Operations Management*, 5th edition (Somerset, NJ: John Wiley & Sons, 2013).

14. R. H. Hayes and S. C. Wheelwright, "Link Manufacturing Process and Product Life Cycles," *Harvard Business Review* (January-February 1979): 133–140.

15. Daniel Swan, Sanjay Pal, and Matt Lippert, "Finding the Perfect Fit," *CSCMP's Supply Chain Quarterly* 4 (2009): 24–32.

16. Kathleen Yeh, "Online and Offline Retailing: Not Just a Science," *The Online Economy: Strategy and Entrepreneurship*, November 13, 2012, http://www.onlineeconomy.org/tag/clientelling.

17. Thomas H. Davenport, "Realizing the Potential of Retail Analytics: Plenty of Food for Those with the Appetite," *Working Knowledge Research Report*, Babson Executive Education White Paper, 2009, pp. 1–42.

18. Ibid.

19. Mary K. Pratt, "How Progressive Uses Telematics and Analytics to Price Car Insurance," *CIO Magazine*, July 26, 2013, www.cio.com/article/736686/How_Progressive_Uses_Telematics_and_Analytics_to_Price_Car_Insurance.

20. Customer Analytics, Cutting a New Path to Growth and Customer Performance, Accenture White Paper, 2010.

21. Emily Steel, "Marketers Watch as Friends Interact Online," *The Wall Street Journal*, April 15, 2010.

22. Ibid.

23. Case Study: Tesco—Pearson. Pearson Education. Available at http://wps.pearsoned.co.uk/ema_uk_he_kotler_euromm_1/126/32286/8265276.cw/content/.

24. Thomas H. Davenport, "Realizing the Potential of Retail Analytics: Plenty of Food for Those with the Appetite," *Working Knowledge Research Report*, Babson Executive Education White Paper, 2009, pp. 1–42.

25. Ibid.

26. Ibid.

27. James A. Cook, "Running Inventory like a Deere," *CSCMP's Supply Chain Quarterly* 4 (2007).

28. Nada Sanders, *The Definitive Guide to Manufacturing and Service Operations* (Lombard, IL: Council of Supply Chain Management Professionals, 2014).

29. J. Maniyaka et al., "Big Data: The Next Frontier for Innovation, Competition, and Productivity," McKinsey Global Institute White Paper, May 2011.

    Basel Kayyali, David Knott, and Steve Van Kuiken, "The Big-Data Revolution in US Health Care: Accelerating Value and Innovation," *McKinsey & Company*, April 2013.

30. Thomas H. Davenport, "Realizing the Potential of Retail Analytics: Plenty of Food for Those with the Appetite," *Working Knowledge Research Report*, Babson Executive Education, 2009, pp. 1–42.

31. Ibid.

32. Ibid.

33. "Best Practices in Customer Service and Store Performance Management," Aberdeen Group White Paper, September 2005.

Chapter 9

1. Thomas H. Davenport, "Competing on Analytics," *Harvard Business Review* (January 2006): 1–9.

2. Thomas H. Davenport, "Realizing the Potential of Retail Analytics: Plenty of Food for Those with the Appetite," *Working Knowledge Research Report*, Babson Executive Education White Paper, 2009, pp. 1–42.

3. Michael Fitzgerald, "How Starbucks Has Gone Digital," *MIT Sloan Management Review*, April 4, 2013.

4. N. R. Sanders and A. Locke, "Making Sense of Outsourcing," *Supply Chain Management Review* 9, no. 2 (2005): 38–45.

5. Ibid.

6. Capabilities, Dunnhumby, available at http://www.dunnhumby.com/us/capabilities

7. Target Provides Update on Data Breach and Financial Performance, Target, January 10, 2014, available at http://pressroom.target.com/news/target-provides-update-on-data-breach-and-financial-performance

8. "Target Data Hack Affected 70 Million People," CBC News, January 20, 2014, http://www.cbc.ca/news/business/target-data-hack-affected-70-million-people-1.2491431.

9. J. Maniyaka et al., "Big Data: The Next Frontier for Innovation, Competition, and Productivity," McKinsey Global Institute White Paper, May 2011.

   Henry Blodget & Alex Cocotas, The State of the Internet, Business Insider, available at http://www.businessinsider.com/state-of-internet-slides-2012-10?op=1

10. Thomas Davenport, "How to Design Smart Business Experiments," *Harvard Business Review* (February 2009).

11. J. Maniyaka et al., "Big Data: The Next Frontier for Innovation, Competition, and Productivity," McKinsey Global Institute White Paper, May 2011.

   Henry Blodget & Alex Cocotas, The State of the Internet, Business Insider, available at http://www.businessinsider.com/state-of-internet-slides-2012-10?op=1

12. Thomas H. Davenport, Jeanne G. Harris, and Robert Morison, *Analytics at Work: Smarter Decisions, Better Results* (Boston, MA: Harvard Business Press, 2010).

13. Thomas H. Davenport and Jeanne G. Harris, *Competing on Analytics: The New Science of Winning* (Cambridge, MA: Harvard Business School Publishing Corporation, 2007).

14. Thomas H. Davenport, "Competing on Analytics," *Harvard Business Review* (January 2006): 1–9.

15. Ibid.

16. Thomas H. Davenport, "Realizing the Potential of Retail Analytics: Plenty of Food for Those with the Appetite," *Working Knowledge Research Report*, Babson Executive Education White Paper, 2009, pp. 1–42.

17. Ibid.

18. Ibid.

19. Aberdeen Group, September 2010.

20. Thomas H. Davenport, "Realizing the Potential of Retail Analytics: Plenty of Food for Those with the Appetite," *Working Knowledge Research Report*, Babson Executive Education White Paper, 2009, pp. 1–42.

21. Andrew McAfee and Erik Brynjolfsson, "Big Data: The Management Revolution," *Harvard Business Review* (October 2012).

22. Jessi Hempel, "IBM's Massive Bet on Watson," *Fortune*, October 7, 2013.

23. Thomas H. Davenport, "Competing on Analytics," *Harvard Business Review* (January 2006): 1–9.

24. J. Maniyaka et al., "Big Data: The Next Frontier for Innovation, Competition, and Productivity," McKinsey Global Institute White Paper, May 2011.

    Henry Blodget & Alex Cocotas, The State of the Internet, Business Insider, available at http://www.businessinsider.com/state-of-internet-slides-2012-10?op=1

25. Thomas H. Davenport, "Competing on Analytics," *Harvard Business Review* (January 2006): 1–9.

26. Derek Dean and Caroline Webb, "Recovering from Information Overload," McKinsey & Company, January 2011, available at http://www.mckinsey.com/insights/organization/recovering_from_information_overload.

Chapter 10

1. Kwame Opam, "Amazon Plans to Ship Your Packages Before You Even Buy Them," *The Verge*, January 18, 2014.

2. S. A. Mohrman and E. E. Lawler, III, "Generating Knowledge That Drives Change," *Academy of Management Perspectives* 26, no. 1(2012): 41–51.

3. J. Shin, M. Taylor, and M. Seo, "Resources for Change: The Relationships of Organizational Inducements and Psychological Resilience to

Employees' Attitudes and Behaviors Toward Organizational Change," *Academy of Management Journal* 55, no. 3 (2012): 727–748.

4. S. A. Mohrman and E. E. Lawler, III, "Generating Knowledge That Drives Change," *Academy of Management Perspectives* 26, no. 1(2012): 41–51.

5. A. H. Van de Ven, and M. S. Poole, (l995). "Explaining Development and Change in Organizations," *Academy of Management Review* 20, no. 3 (1995): 510–540.

6. M. H. Kavanagh and N. M. Ashkansasy, "The Impact of Leadership and Change Management Strategy on Organizational Culture and Individual Acceptance of Change During a Merger," *British Journal of Management* 17 (2006): 81–103.

7. V. Garcia-Morales, M. Jimenez-Barrionuevo, and L. Gutierrez-Gutierrez, "Transformational Leadership Influence on Organizational Performance Through Organizational Learning and Innovation," *Journal of Business Research* 65, no. 7 (2012): 1040–1050.

8. Steven Harner, "Can Rakuten Maverick CEO Mikitani's New Organization Change Japan?" *Forbes*, October 10, 2012.

9. Tsedal Neeley, "Global Business Speaks English," *Harvard Business Review* (May 2012).

10. Lewin, K., *Field Theory in Social Science; Selected Theoretical Papers*. D. Cartwright (ed.) (New York: Harper & Row, 1951); M. Crossan, H. W. Lane, and R. E. White, "An Organizational Learning Framework: From Intuition to Institution," *Academy of Management Review* 24, no. 3 (1999): 522–537; B. Burnes, "Kurt Lewin and the Planned Approach to Change: A Re-appraisal," *Journal of Management Studies* 41, no. 6 (September 2004).

11. B. Michael, M. J. Neubert, and R. Michael, "Three Alternatives to Organizational Value Change and Formation: Top-Down, Spontaneous Decentralized, and Interactive Dialogical," *Journal of Applied Behavioral Science* 48, no. 3 (2012): 380–409.

"Online Extra: The Secret to BMW's Success," *Bloomberg Businessweek*, October 15, 2006.

12. *Chapter 25: Regeneration Through Team Spirit*, Sony Corporate Info, available at http://www.sony.net/SonyInfo/CorporateInfo/History/SonyHistory/2-25.html

13. R. A. Eisenstat, M. Beer, N. Foote, T. Fredberg, and F. Norrgren, "The Uncompromising Leader," *Harvard Business Review* 86, no. 7/8 (2008): 50–57.

14. Carol Nobel, "How IT Shapes Top Down and Bottom Up Decision Making," *Harvard Business Review* (November 1, 2010), hbswk.hbs.edu/item/6504.html.

15. Victor Mayer-Schönberger and Kenneth Cukier, *Big Data, A Revolution That Will Transform How We Live, Work, and Think* (Boston, MA: Houghton Mifflin Harcourt, 2013, p. 135–136).

16. Ibid.

17. Peter Katel, "Bordering on Chaos," *Wired*, March 14, 2012, pp. 1–6.

18. Mitchell J. Neubert and Bruno Dyck, *Organizational Behavior* (Somerset, NJ: John Wiley & Sons, 2014).

19. O. Muurlink, A. Wilkinson, D. Peetz, and K. Townsend, "Managerial Autism: Threat-Rigidity and Rigidity's Threat," *British Journal of Management* 23, suppl. 1 (2012): S74–S87; B. Dyck, (1996). "The Role of Crises and Opportunities in Organizational Change," *Non-Profit and Voluntary Sector Quarterly* 25 (1996): 321–346.

20. M. Fugate, G. E. Prussia, and A. J. Kinicki, "Managing Employee Withdrawal During Organizational Change: The Role of Threat Appraisal," *Journal of Management* 38, no. 3 (2012): 890–914.

21. A. A. Armenakis and S. G. Harris, "Reflections: Our Journey in Organizational Change Research and Practice," *Journal of Change Management* 9, no. 2 (2009): 127–142.

22. T. Peters and R. J. Waterman, *In Search of Excellence* (New York, NY: HarperCollins, 2004).

23. Mitchell J. Neubert and Bruno Dyck, *Organizational Behavior* (Somerset, NJ: John Wiley & Sons, 2014).

24. Ibid.

25. E. McKirt, "Artist. Athlete. CEO.," *Fast Company*, 114 (September 2010): 66–74, http://www.fastcompany.com/1676902/how-nikes-ceo-shook-shoe-industry.

26. James B. Stewart, Looking for a Lesson in Google's Perks, The New York Times, March 15, 2013, available at http://www.nytimes.com/2013/03/16/business/at-google-a-place-to-work-and-play.html?pagewanted=all&_r=0

27. Ibid.

28. Thomas H. Davenport and Jerry O'Dwyer, "Tap Into the Power of Analytics," *CSCMP's Supply Chain Quarterly* 4 (2011).

29. Jessi Hempel, "IBM's Massive Bet on Watson," *Fortune*, October 7, 2013.

30. Mary Siegfried, "Find the Big Picture in Bit Data," *Inside Supply Management Magazine* (January-February 2014): 23.

31. Mitchell J. Neubert and Bruno Dyck, *Organizational Behavior* (Hoboken, New Jersey: John Wiley & Sons, 2014).

32. Thomas H. Davenport, "Realizing the Potential of Retail Analytics: Plenty of Food for Those with the Appetite," *Working Knowledge Research Report*, Babson Executive Education White Paper, 2009, pp. 1–42.

33. James Cook, "From Many, One: IBM's Unified Supply Chain," *CSC-MP Supply Chain Quarterly* 4 (2012).

34. Thomas H. Davenport, "Competing on Analytics," *Harvard Business Review* (January 2006): 1–9.

Index